# Airport Ground Video Surveillance

Xiang Zhang • Honggang Wu • Guoqiang Wang •
Jian Cheng

# Airport Ground Video Surveillance

## Algorithms and Applications

Xiang Zhang
School of ICE
UESTC
Chengdu, China

Honggang Wu
Research and Development Center
The Second Research Institute of CAAC
Chengdu, China

Guoqiang Wang
Research and Development Center
The Second Research Institute of CAAC
Chengdu, China

Jian Cheng
School of ICE
UESTC
Chengdu, China

ISBN 978-981-96-2309-9     ISBN 978-981-96-2310-5   (eBook)
https://doi.org/10.1007/978-981-96-2310-5

This Springer imprint is published by the registered company Springer Nature Singapore Pte Ltd.
The registered company address is: 152 Beach Road, #21-01/04 Gateway East, Singapore 189721,
Singapore

If disposing of this product, please recycle the paper.

*To my beloved family, Peng Yan, Zhi Yu, and Zhi Ling*
*To my dear students, Yang Wentao*

# Preface

As air passenger and cargo volumes continue to grow rapidly, the global civil aviation industry is booming. The upward trajectory in air traffic adds the most pressure on airports. The airport ground area is an important space for aircraft to take off, land, taxi, and park. Because airports are becoming increasingly crowded, operational efficiency is low, leading to frequent safety incidents. For example, on January 2, 2024, two planes collided and caught fire at Haneda Airport, resulting in the death of five people; on January 16, two planes collided at New Chitose Airport in Hokkaido; on February 1, two planes collided again at Osaka Itami Airport.

Airports must evolve to become smarter in order to tackle efficiency and safety challenges. A crucial aspect of intelligent airports is airport ground video surveillance, which involves the perception and analysis of ground operating conditions using computer vision technology. This method addresses several limitations of manual surveillance, such as blind spots, personnel fatigue, and high labor costs. Consequently, airport ground video surveillance can assist or even partially replace airport management personnel, significantly enhancing the intelligence level of the airport.

Airport ground video surveillance encompasses two main components: applications and algorithms. Typical applications include visual conflict alarms, visual docking guidance, and augmented reality. For instance, the augmented reality application transmits real-time airport video to the control center, analyzes the status of aircraft in the footage, and overlays relevant data, such as flight numbers, onto the video. This real-time display aids air traffic controllers in accurately and promptly assessing the airport situation. The foundation of these intelligent applications lies in various computer vision algorithms, including target recognition, segmentation, and tracking. However, the performance of these algorithms often declines when applied to airport environments compared to laboratory settings. For example, our research indicates that the accuracy of tracking algorithms can drop by as much as 40% when transitioning from laboratory scenes to airport scenes. Therefore, it is essential to develop specialized computer vision algorithms tailored for airport environments, which can then be used to create various airport ground video surveillance applications.

This book provides a comprehensive overview of current research in airport ground surveillance, organized into four key sections: Background, Dataset, Algorithm, and Application. In the Background section, we define intelligent airports and airport ground video surveillance and analyze the significance of the latter within intelligent airports. The Dataset section highlights publicly available image and video datasets for airport scenes, including the Airport Ground Video Surveillance (AGVS) dataset series developed by our team. In the Algorithm section, we discuss the design principles of computer vision algorithms specifically for airport ground settings, as well as several notable algorithms for airport ground video surveillance. Unlike algorithms in fundamental research, which prioritize generalization, those designed for airport scenarios focus on achieving optimal performance in specific contexts. To this end, airport ground video surveillance algorithms often leverage unique prior information relevant to airport environments to enhance performance. While such algorithms may lack generalizability due to their reliance on specific airport data, they are better aligned with the operational needs of intelligent airports. Based on these principles, we have developed a series of algorithms for segmentation, recognition, and tracking, which will be detailed in the Algorithm section. In the Application section, we will outline the design methods for airport ground video surveillance applications and present a typical applications developed by our team, the Airport Panoramic Enhanced Surveillance (APES) system. Finally, we summarize the key findings of this book and look forward to future research directions in airport ground surveillance.

This book is the first of its kind to comprehensively summarize research on airport ground video surveillance. It provides detailed insights from airport ground datasets to algorithm research and the design of intelligent airport applications. This resource is valuable not only for airport management and research personnel but also for scholars engaged in fundamental computer vision research.

Chengdu, China                                                                          Xiang Zhang
                                                                                                    Honggang Wu
                                                                                                    Guoqiang Wang
                                                                                                         Jian Cheng

# Acknowledgments

We would like to express our gratitude to Wentao Yang for his exceptional efforts in compiling this book. Additionally, we extend our thanks to Professor Xiang Zhang's students—Maozhang Zhou, Youhen Xiao, Wei Lin, Yonghui Huang, Haixin Zeng, and Runzhi Gao—for their valuable contributions. Lastly, we appreciate Professor Xiaofeng Li for his thoughtful guidance throughout the process of creating this book. This project is also supported by the National Natural Science Foundation of China (Grant No. U1733111).

# Contents

# Chapter 1
# Background Introduction

**Abstract** The civil aviation industry is a vast and intricate system that plays a crucial role in the global transportation network. In this chapter, we will first provide an overview of the current status of the global civil aviation industry, followed by an analysis of the role of airports within this sector. Next, we will introduce the concept of intelligent airports, explore the motivations behind their development, and discuss the significant role of airport ground surveillance in enhancing the functionality of intelligent airports.

## 1.1 Civil Aviation

The civil aviation industry is a vital component of the global transportation system, significantly influencing production activities and social interactions while playing an essential role in the modern economy. It connects cities across countries and regions through airlines, airports, and routes. As a key player in global passenger transportation, civil aviation enables quick travel and supports the tourism sector in many areas. Additionally, it is crucial for global trade, particularly for transporting high-value and time-sensitive goods. Furthermore, the industry not only generates numerous direct jobs but also stimulates related sectors such as tourism, hospitality, and catering.

The globalization of the civil aviation industry has encouraged countries to collaborate on aviation policies and regulations and to establish safety standards, environmental policies, and air rights agreements through organizations like the International Civil Aviation Organization (ICAO) to foster a healthy global aviation market. However, the rapid growth of the industry has led to environmental challenges, particularly carbon emissions. The sector is exploring sustainable aviation technologies to mitigate its environmental impact while partnering with other transportation modes to tackle the challenges of global climate change.

X. Zhang et al., *Airport Ground Video Surveillance*,
https://doi.org/10.1007/978-981-96-2310-5_1

In 2019, the International Air Transport Association (IATA) reported global passenger traffic at 4.54 billion. Despite the significant impact of the COVID-19 pandemic, demand in the aviation market continued to grow in 2023, with full recovery anticipated by 2024. The Asia-Pacific region, particularly East Asia, has rebounded more quickly, exemplified by China's domestic aviation market, which saw a year-on-year growth of 138.8% in 2023.

## 1.2    Airport

The civil aviation industry operates through collaboration among airlines, airports, air traffic control, ground services, and maintenance teams. The process for a civil aviation flight begins with passengers purchasing tickets online or offline. On flight day, airline and airport staff perform a takeoff check, followed by security checks and boarding. Once airborne, the pilot navigates until landing at the destination, where the plane taxis to the apron for passenger disembarkation and luggage retrieval. After each flight, the airline conducts routine maintenance and inspections to ensure safety and efficiency. Airports are crucial to this system, facilitating all activities except for flights. Thus, the airport's operational efficiency and service quality significantly impact the overall performance of the civil aviation industry and passenger experience.

There are over 10,000 airports worldwide, accommodating approximately 100,000 flights daily, with 30% being international and 70% domestic. Major international hubs like Hartsfield-Jackson Atlanta, Beijing Capital, and London Heathrow see annual passenger traffic exceeding 50 million. Global air cargo volume typically ranges from 50 to 60 million tons. Airports can be categorized by their functions: International airports handle international flights and feature comprehensive customs and security; domestic airports are smaller and cater to domestic routes; regional airports connect specific areas with local and short-haul services; military airports serve military operations but may also host civilian flights; and cargo airports specialize in freight transport with efficient handling facilities. General aviation airports support private and small airline operations.

Despite the variety of airport types, they share similar infrastructure and layouts. All airports consist of terminals, aprons, runways, and taxiways, as illustrated in Fig. 1.1. The area where aircraft move—comprising aprons, runways, and taxiways—is known as the airport ground. Although the airport looks spacious, large cities usually have only one or two international airports, leading to crowded conditions as all aircraft must operate within the limited space of the airport ground.

**Fig. 1.1** Illustration of the airport structure, which generally includes the terminal, apron, and runways-taxiway area. All airports have a similar structure

## 1.3   Airport Ground Video Surveillance for Intelligent Airport

The rapid growth of civil aviation traffic has made improving airport efficiency and safety an urgent priority. In response, the concept of the intelligent airport has emerged. These airports leverage advanced technologies like the Internet of Things, artificial intelligence, big data, and cloud computing to foster a safer, more convenient, and efficient environment, enabling intelligent management of operations and promoting sustainable development.

The airport ground, that is, the area where aircraft take off, land, taxi and park, as shown in Fig. 1.1, is closely related to aircraft safety, which can be said to be the top priority of airport management. However, with the rapid development of the civil aviation industry, the airport ground has become increasingly crowded, resulting in various safety accidents occurring from time to time. Therefore, airport ground surveillance is crucial to ensure airport safety. In the past, airport managers either visually monitored the airport in the observation room of the airport tower or conducted on-site inspections of key areas such as the airport runway every few hours, as shown in Fig. 1.2. However, surveillance using the naked eye has notable disadvantages, including blind spots, personnel fatigue, and high costs. In today's intelligent airports, strategically placed cameras can monitor the airport's operations and send data to the airport tower or a remote monitoring center. This enables airport managers to gain real-time insights into overall operations, as illustrated in Fig. 1.3. Additionally, the airport ground video surveillance system employs computer vision technology to analyze video data, facilitating various intelligent functions.

**Fig. 1.2**  Illustration of airport ground surveillance based on the naked eye

**Fig. 1.3**  Illustration of airport ground video surveillance based on cameras

Camera-based airport ground surveillance, or airport ground video surveillance, is an important part of intelligent airports. Airport ground video surveillance includes two parts: algorithms and applications. In the application part, there are various intelligent applications such as the airport panorama application, the visual conflict alarm application, the visual docking guidance application, etc. And these applications are all based on basic computer vision algorithms such as object recognition, segmentation, and tracking. However, our experiments show that the recognition, segmentation, and tracking algorithms in theoretical research have a sharp drop in performance in airport scenarios, sometimes even more than 40%. This is because the airport is a special scene, which is very different from the scenes commonly seen in theoretical research data, such as street intersections, squares, and halls. For example, the monitoring distance in urban scenes is generally no more than 100 meters, while the monitoring distance at an airport can reach several thousand meters. Therefore, the state-of-the-art algorithm in theoretical research cannot be directly applied to airport ground surveillance but must develop dedicated computer vision algorithms for the special challenges faced by airport scenarios and on this basis develop various airport ground surveillance applications. Airport recognition, segmentation, and tracking are illustrated in Fig. 1.4.

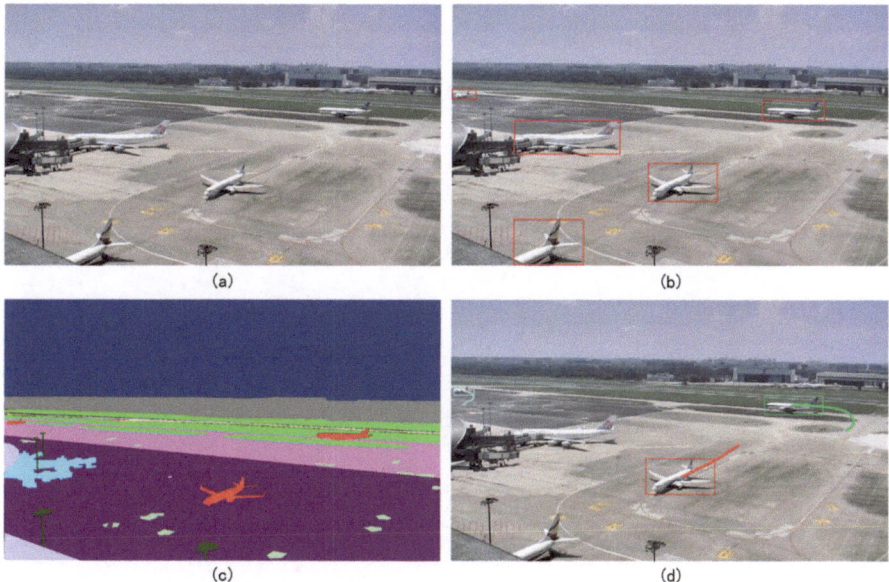

**Fig. 1.4** (**a**) A frame. (**b**)–(**d**) Illustration of airport recognition, segmentation, and tracking

Research on computer vision for airport ground video surveillance is limited, primarily due to a lack of dedicated surveillance datasets. This book summarizes existing studies, focusing on the collaborative efforts of the University of Electronic Science and Technology of China and the Second Research Institute of the Civil Aviation Administration of China. Their collaboration covers three main areas: datasets, algorithms, and applications for airport ground video surveillance.

# Chapter 2
# Datasets

**Abstract** In this chapter, some public datasets related to airport ground video surveillance are introduced. They are AGVS-S, AGVS-SS, AGVS-R, AGVS-T, AGVS-AR, FGVC-Aircraft, and ALERT. The first five datasets belong to the AGVS series that are proposed by our team, and before this there were no dedicated datasets for airport ground video surveillance. The AGVS series encompasses the three core research areas of computer vision—segmentation, recognition, and tracking—thus facilitating research on various airport ground computer vision topics. AGVS-S focuses on change detection, segmenting video pixels into moving foreground and background. AGVS-SS provides semantic segmentation by classifying each pixel into predefined classes. AGVS-R is dedicated to image recognition of moving targets on the ground. AGVS-T is a tracking dataset that highlights unique aircraft motion patterns and related challenges. AGVS-AR involves the ongoing production of an aircraft action recognition dataset. Although FGVC-Aircraft and ALERT do not specifically address airport ground issues, they are connected to aircraft and airport research.

## 2.1 AGVS-S Dataset

### 2.1.1 Design and Challenges

The AGVS-S dataset was specifically designed for change detection in airport ground. Change detection is a binary segmentation problem, which aims to classify pixels into two classes, moving foreground and background. AGVS-S contains 25 long video sequences with approximately 100,000 frames. Each frame is annotated with precise pixel-level ground truth, as shown in Fig. 2.1. The dataset covers a variety of challenges unique to airport environments, which are described as follows.

**Haze** Haze is a natural phenomenon that diminishes image contrast, making it harder to distinguish between the foreground and background. It occurs more frequently in airport videos due to the increased monitoring distance, which amplifies the effects of haze. Even slight haze becomes apparent at longer distances,

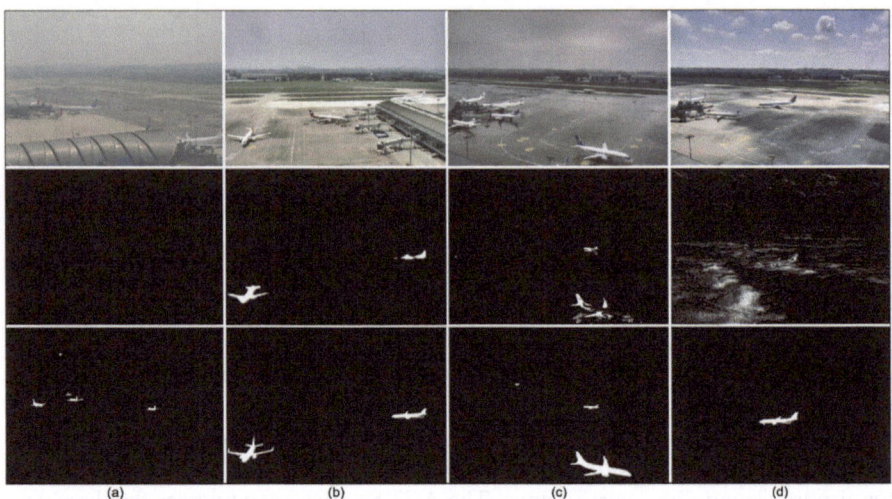

**Fig. 2.1** The AGVS-S dataset. The first row are typical frames from AGVS-S, and the last row are the pixel-level ground truth. The second row shows segmentation results by four classic change detection algorithms. Reprinted with permission from [1]

typically several kilometers at airports, leading to significant impact. As illustrated in Fig. 2.1a, the algorithm fails to detect all moving targets under moderate haze conditions.

**Camouflage** Camouflage becomes challenging when a moving object matches the color distribution of its background, a frequent issue at airports where gray-white concrete and predominantly white aircraft blend together. This effect, akin to haze, diminishes the contrast between the foreground and background, leading to significant detection failures. Camouflage is a major obstacle in change detection, as evidenced by the examples in Fig. 2.1, which all exhibit some level of camouflage.

**Simultaneous Multi-scale Detection** Because of the extremely long surveillance distance of the airport, objects at different distances are usually captured in one frame. This distance difference is very large, ranging from tens of meters to thousands of meters, allowing everything from detailed close-ups to blurred distant views to appear at the same time. The AGVS-S dataset named this phenomenon simultaneous multi-scale. Simultaneous multi-scale poses a huge challenge to change detection because they must detect objects of different scales at the same time, while the processing methods of objects of different scales are sometimes incompatible. The simultaneous multi-scale phenomenon is very obvious in Fig. 2.1.

**Shadow and Nonuniform Illumination** Airports are exposed outdoor environments that are highly affected by changes in shadows and illumination. Throughout the day, natural illumination shifts from morning to noon and from dusk to late night, resulting in uniform grayscale variations in imaging. Additionally, moving aircraft and clouds can create rapidly changing, nonuniform shadows on the ground, as illustrated in Fig. 2.1d. These illumination changes can trigger widespread false alarms.

**Shape and Color Variation** Aircraft are irregularly shaped objects with varying wingspans, making segmentation, recognition, and tracking particularly challenging. Their shapes significantly alter with changes in perspective, and different designs and paint schemes lead to further variability in appearance. This diversity poses a substantial challenge.

**Long Strips** The AGVS-S dataset treats long strips as a distinct challenge due to the slender fuselage and wings of targets like civil airliners, which are central to airport ground video surveillance. Long objects can result in incomplete or broken detections, making it essential to ensure detection integrity for applications such as visual docking guidance and conflict alerts.

**PTZ Camera** The use of Pan-Tilt-Zoom (PTZ) cameras adds complexity as both foreground and background move, challenging unsupervised change detection methods that usually assume a stationary background. The last three sequences in AGVS-S are captured with PTZ cameras.

The AGVS-S dataset also contains some other challenges, such as weather changes in Fig. 2.1c. From the AGVS-S website, we can see that there are six videos that do not provide ground truth. These six videos contain some extreme situations, such as extremely low illumination in the middle of the night with bright halos caused by strong lighting.

## 2.1.2 Data Collection and Annotation

Through our long-term partnership with the Second Institute of Civil Aviation Administration of China, the University of Electronic Science and Technology of China obtained permission to collect raw data from an airport in southwest China. We selected 25 diverse segments from the extensive raw data to create the AGVS-S dataset, as illustrated in Fig. 2.2. The subsequent step is data annotation, which is also depicted in Fig. 2.3. Key principles to ensure the accuracy and reliability of the annotation are outlined below.

**Confirmation of Motion Status** AGVS-S is a change detection dataset that requires annotation only for moving objects. Objects that are stationary or that have not yet started moving do not need annotation. To accurately annotate only moving objects, the annotator must repeatedly watch the entire video.

**Size Consideration** Because there are many very small objects in the airport scene video that are difficult to distinguish, an important question is how big an object should be annotated, and how small an object should not be annotated? Our principle is that objects that are too small to be distinguished by the naked eye should not be

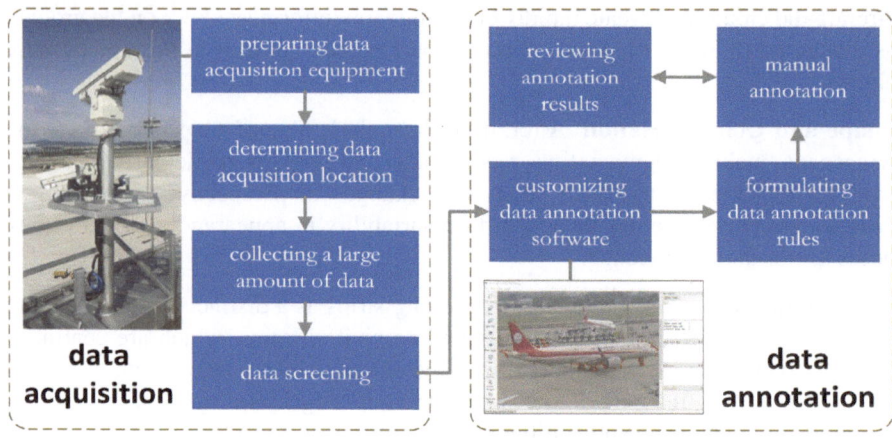

**Fig. 2.2**  Data acquisition and annotation flowchart. Reprinted with permission from [1]

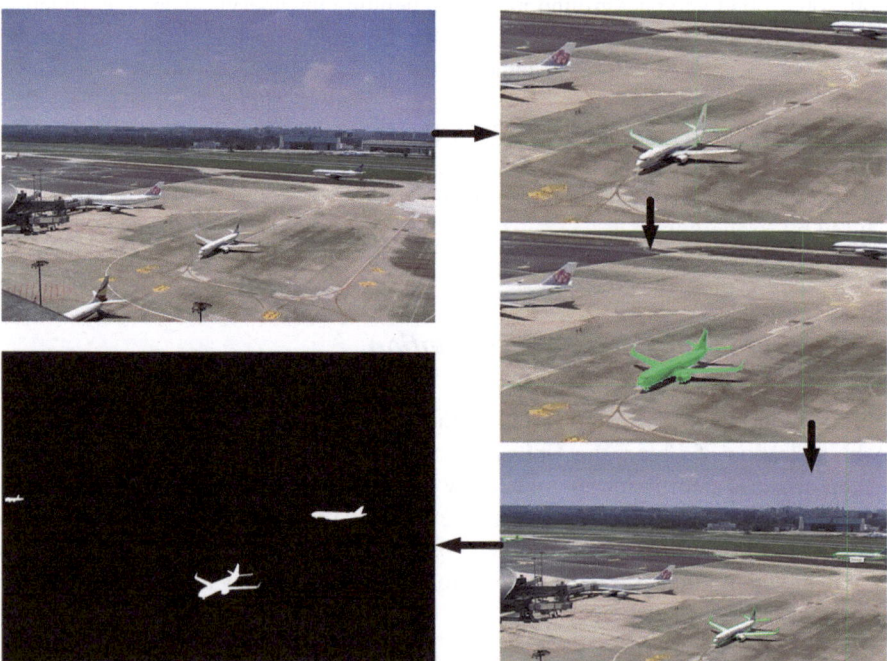

**Fig. 2.3**  Annotation process of AGVS-S dataset. Adapted with permission from [1]

annotated; otherwise they must be annotated. During the data annotation process, we also found that the objects that do not need to be annotated are generally less than 100 pixels in size. Therefore, 100 pixels is an important empirical value.

**Accuracy Checks** Due to the subjectivity of manual annotation, strict review guidelines are necessary to ensure accuracy. To tackle this issue, we have established a multi-person review process where annotations are independently evaluated by several reviewers. If all reviewers agree on the accuracy, the annotation is approved. If any reviewer identifies a problem, their feedback is returned to the annotator for reevaluation, continuing until consensus is reached.

The annotation tool for the AGVS-S dataset is a customized version of LabelImg, enhanced with features to improve annotation efficiency. One such feature allows users to load a mask from the previous frame and overlay it on the current frame, enabling annotators to achieve results by merely fine-tuning the loaded mask. This customized tool is available for download on the AGVS-S dataset website.

The process of annotating a frame using custom software is shown in Fig. 2.3. The annotator loads and observes the image, zooms in on the area containing a moving aircraft, and draws an accurate polygon mask along the aircraft outline. The annotator then repeats the above steps for all moving objects in the image and finally obtains the complete ground truth of this frame. To ensure accuracy, two independent reviewers will carefully review the annotation results. Only when both reviewers agree can the next frame be annotated; otherwise it will be re-annotated.

### *2.1.3   Comparative Experiments*

Experiments are conducted on AGVS-S to compare existing change detection algorithms. Totally 21 state-of-the-art algorithms are tested in [1]. $RE$, $PR$, and $F-Measure$ ($FM$) are chosen for quantitative comparison of detection accuracy,

$$RE = \frac{TP}{TP+FN},$$

$$PR = \frac{TP}{TP+FP},$$

$$FM = \frac{2 \times RE \times PR}{RE+PR},$$

where $TP$, $FP$, and $FN$ represent the counts of true positives, false positives, and false negatives, respectively. The quantitative results of five comparison algorithms are shown in Table 2.1.

The AGVS-S dataset contains 25 sequences, but only the results of the first 22 sequences are given in Table 2.1. This is because the last three sequences are based on the PTZ camera, and the unsupervised algorithm cannot handle this situation. As can be seen from Table 2.1, SuBSENSE [2] and CascadeCNN [5] are the best unsupervised and supervised algorithms, respectively, and the performance of CascadeCNN is much better than SuBSENSE. However, this does not mean that the unsupervised algorithm has lost its research significance, because the training set and test set of the supervised algorithm in Table 2.1 are not separated, that is,

**Table 2.1** Performance comparison of three unsupervised algorithms and two supervised algorithms on AGVS-S. Reprinted with permission from [1]

| Sequences | SuBSENSE [2] | | | ViBe [3] | | | Codebook [4] | | | CascadeCNN [5] | | | FgSegNet [6] | | |
|---|---|---|---|---|---|---|---|---|---|---|---|---|---|---|---|
| | RE | PR | FM | RE | PR | FM | RE | PR | FM | RE | PR | FM | RE | PR | FM |
| S1 | 0.55 | 0.54 | 0.55 | 0.33 | 0.43 | 0.37 | 0.61 | 0.43 | 0.50 | **0.97** | 0.70 | **0.82** | 0.88 | **0.74** | 0.81 |
| S2 | 0.73 | 0.68 | 0.71 | 0.44 | 0.66 | 0.53 | 0.70 | 0.03 | 0.06 | **0.98** | 0.87 | 0.92 | 0.96 | **0.97** | **0.97** |
| S3 | 0.70 | 0.74 | 0.72 | 0.47 | 0.74 | 0.58 | 0.60 | 0.10 | 0.18 | **0.99** | 0.97 | 0.98 | 0.98 | **0.98** | **0.98** |
| S4 | 0.43 | 0.60 | 0.50 | 0.29 | 0.46 | 0.36 | 0.49 | 0.06 | 0.10 | **0.99** | 0.95 | **0.97** | 0.94 | **0.97** | 0.96 |
| S5 | 0.62 | 0.35 | 0.44 | 0.48 | 0.45 | 0.46 | 0.71 | 0.28 | 0.40 | 0.96 | **0.96** | **0.96** | **0.98** | 0.92 | 0.95 |
| S6 | 0.69 | 0.63 | 0.66 | 0.57 | 0.58 | 0.58 | 0.74 | 0.51 | 0.61 | **0.94** | 0.85 | 0.89 | 0.90 | **0.94** | **0.92** |
| S7 | 0.73 | 0.51 | 0.60 | 0.55 | 0.62 | 0.59 | 0.73 | 0.18 | 0.29 | **0.97** | 0.81 | 0.88 | 0.98 | **0.92** | **0.95** |
| S8 | 0.64 | 0.46 | 0.54 | 0.61 | 0.51 | 0.56 | 0.72 | 0.33 | 0.45 | **0.96** | 0.71 | 0.82 | 0.89 | **0.88** | **0.88** |
| S9 | 0.61 | 0.50 | 0.55 | 0.58 | 0.57 | 0.57 | 0.67 | 0.19 | 0.30 | **0.97** | 0.85 | 0.91 | 0.94 | **0.96** | **0.95** |
| S10 | 0.52 | 0.74 | 0.61 | 0.53 | 0.63 | 0.58 | 0.77 | 0.43 | 0.55 | **0.98** | 0.91 | 0.94 | 0.97 | **0.93** | **0.95** |
| S11 | 0.47 | 0.71 | 0.56 | 0.44 | 0.61 | 0.51 | 0.58 | 0.08 | 0.14 | **0.97** | 0.91 | 0.94 | 0.97 | **0.93** | **0.95** |
| S12 | 0.44 | 0.57 | 0.49 | 0.50 | 0.50 | 0.58 | 0.41 | 0.53 | 0.96 | 0.88 | **0.91** | 0.88 | **0.92** | 0.90 | **0.90** |
| S13 | 0.69 | 0.66 | 0.60 | 0.61 | 0.61 | 0.58 | 0.26 | 0.38 | 0.93 | 0.82 | **0.89** | **0.93** | **0.96** | **0.93** | 0.90 |
| S14 | 0.35 | 0.46 | 0.40 | 0.43 | 0.53 | 0.47 | 0.44 | 0.42 | 0.43 | 0.83 | 0.90 | 0.91 | **0.97** | **0.93** | **0.93** |
| S15 | 0.64 | 0.59 | 0.61 | 0.44 | 0.46 | 0.45 | 0.77 | 0.38 | 0.51 | **0.99** | 0.93 | 0.96 | 0.97 | **0.97** | **0.97** |
| S16 | 0.52 | 0.87 | 0.65 | 0.22 | 0.91 | 0.36 | 0.54 | 0.71 | 0.61 | **0.97** | **0.97** | **0.97** | 0.97 | 0.90 | 0.97 |
| S17 | 0.24 | **0.89** | 0.38 | 0.09 | 0.83 | 0.16 | 0.20 | 0.19 | **0.97** | **0.97** | 0.77 | 0.86 | 0.94 | 0.87 | 0.90 |
| S18 | 0.01 | 0.32 | 0.01 | 0.02 | 0.67 | 0.03 | 0.19 | 0.20 | 0.19 | **0.97** | 0.77 | 0.86 | 0.94 | **0.87** | **0.90** |
| S19 | 0.42 | 0.79 | 0.55 | 0.29 | 0.83 | 0.43 | 0.47 | 0.59 | 0.52 | **0.99** | 0.95 | 0.97 | 0.96 | **0.97** | **0.97** |
| S20 | 0.58 | 0.77 | 0.60 | 0.34 | 0.45 | 0.47 | 0.72 | 0.96 | **0.97** | 0.96 | **0.97** | 0.96 | 0.87 | 0.93 | 0.95 |
| S21 | 0.56 | 0.73 | 0.63 | 0.33 | 0.64 | 0.44 | 0.56 | 0.97 | 0.97 | **0.99** | **0.99** | **0.99** | 0.99 | 0.99 | 0.99 |
| S22 | 0.58 | 0.65 | 0.62 | 0.14 | 0.84 | 0.24 | 0.45 | 0.63 | 0.52 | **0.99** | **0.99** | **0.99** | 0.99 | 0.99 | 0.99 |

The best values are in bold

the first few hundred frames of a sequence are used for training, and the remaining frames of the same sequence are used for testing. In this case, the algorithm is over-fitted. If the test scene changes, even a small change, the algorithm performance may drop sharply. In fact, before the supervised algorithm has good generalization ability, the unsupervised algorithm can be used in practice.

Please note that we initially named the first dataset AGVS-S simply AGVS. After proposing additional datasets like AGVS-T and AGVS-R, we decided to standardize the naming convention for the AGVS series, renaming the change detection dataset to AGVS-S. All datasets in the AGVS series can be found on our website www. agvs-caac.com.

## 2.2 AGVS-SS Dataset

### 2.2.1 Design and Challenges

AGVS-SS is a segmentation dataset for airport ground video surveillance developed by our group. Unlike AGVS-S, which focuses on change detection with two categories, moving targets and background, AGVS-SS targets semantic segmentation, a multi-classification problem that identifies distinct semantic entities in images. Thus, AGVS-SS is designed to encompass a range of semantic categories and challenging scenarios specific to airport ground contexts.

Based on a large number of observations, we have identified more than 10 semantic categories that appear in the airport ground area. They are airplane, boarding bridge car, tractor, airport terminal, boarding bridge, background, signs, landmarks, etc. Some of these semantic categories are shown in Fig. 2.4. In fact, many other semantic categories can be observed in airport videos, such as flying birds, foreign objects, etc. However, these semantic categories are either occasional or difficult to define their semantic attributes, so they are not included in AGVS-SS for the time being. In the future, the dataset can be further expanded or refined as needed.

AGVS-SS has similar challenges as AGVS-S, such as multi-scale, illumination variations and weather variations. However, because semantic segmentation is a multi-classification problem, AGVS-SS also includes some new challenges, such as intra-class variability and inter-domain differences. In general, semantic segmentation is more challenging than change detection, and the performance of existing semantic segmentation methods on AGVS-SS is far from meeting the needs of practical applications.

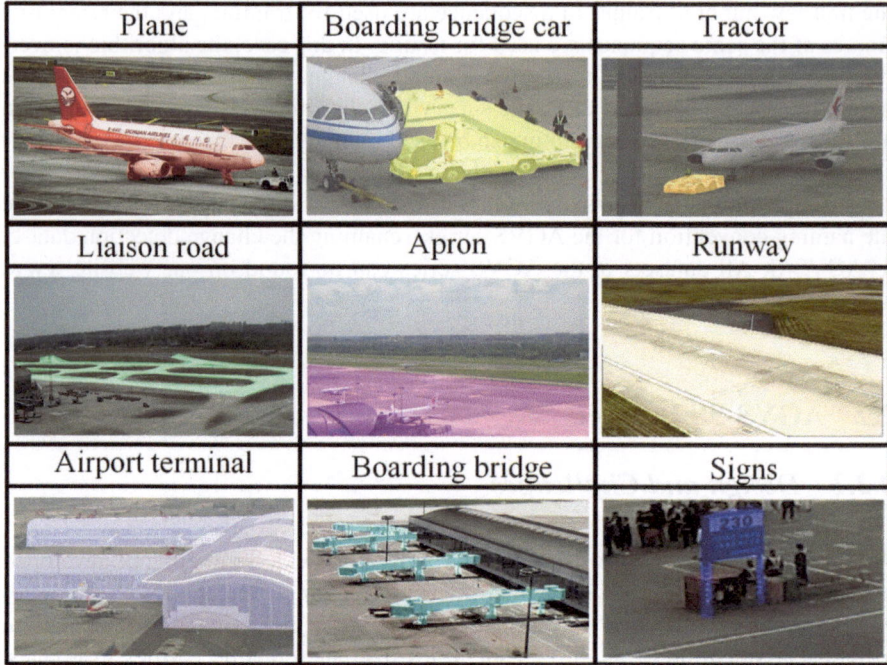

**Fig. 2.4** Illustration of some semantic classes in airport ground. Manually annotated masks are overlaid on top of the semantic entities

## 2.2.2  Data Collection and Annotation

Data was collected and screened similarly to the AGVS-S dataset, resulting in 100 short videos averaging 12 seconds each at 30 FPS, totaling around 80,000 frames currently. Each frame was manually annotated at the pixel level using AnyLabeling software for polygon annotation, alongside adjacent frame reuse and model-assisted labeling functions. We implemented strict annotation rules and review processes to ensure accuracy. The annotation workload for AGVS-SS is significantly larger than that of the AGVS-S dataset due to the requirement for labeling all categories.

## 2.2.3  Experiments

**Experimental Settings** We fine-tuned 11 state-of-the-art semantic segmentation models and tested their performance on AGVS-SS. The 11 algorithms are FCN, PSPNet, Deeplabv3 [7], Deeplabv3+ [8], OCRNet [9], HRNet [10], BiseNetv1 [11], BiseNetv2 [12], SETR [13], SegFormer [14], and Segmentor [15]. The initial learning rate was set to $1 \times 10^{-4}$, the learning rate adjustment strategy was "poly,"

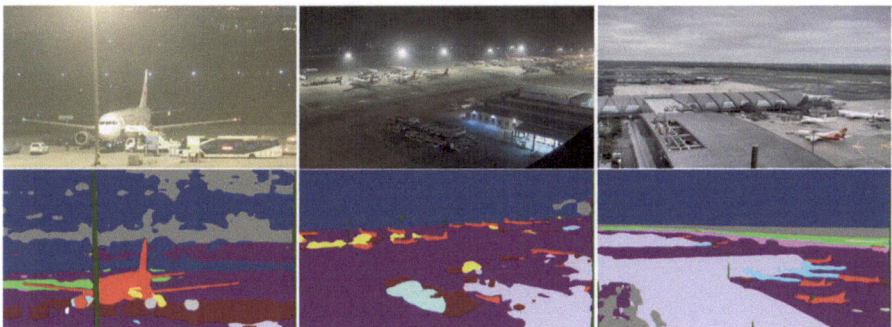

**Fig. 2.5**  Demonstration of semantic segmentation on AGVS-SS

the weight decay coefficient was $5 \times 10^{-4}$, fine-tuning was conducted for 20k epochs, the batch size was set to 16, and the loss function used was cross-entropy loss. Some visual results of semantic segmentation on AGVS-SS are shown in Fig. 2.5.

**Experimental Results and Discussions** IoU is a commonly used evaluation metric for semantic segmentation. IoU measures the overlap between the predicted segmentation results and the ground truth. The IoU of 11 models on AGVS-SS is shown in Table 2.2. It can be seen that the averaged IoU of nine classes on AGVS-SS was less than 50%, and the performance in some categories is very poor. For example, the averaged IoU on Runway class and Person class is even much lower than 10%. Person is a common semantic category, and good semantic segmentation results can often be obtained on other person datasets. However, the averaged IoU of Person on AGVS-SS is only 6.38%. This indicates that the airport ground is indeed a difficult scene, and a large part of the reason is that the monitoring distance of airports is too far, resulting in generally smaller targets.

Please note that the AGVS-SS dataset is still in production, but a preliminary version has been posted on the AGVS website. We will continue to improve the dataset, including expanding the amount of data, the number of categories, and improving the quality of annotation. We will continue to update the dataset.

## 2.3  AGVS-R Dataset

### 2.3.1  Design and Challenges

Object recognition is a critical task in computer vision, which led us to create the AGVS-R recognition dataset for airport ground video surveillance. Unlike other datasets, AGVS-R focuses on the movable targets on airport ground such as the aircraft, as they are central to airport operations. The design principle of AGVS-

**Table 2.2** The IoU of 11 image semantic segmentation models on AGVS-SS. Best scores are marking in bold

| Models | Classes | | | | | | | | | | | |
|---|---|---|---|---|---|---|---|---|---|---|---|---|
| | Sky | Liasion Road | Lawn | Apron | Runway | Signs | Landmark | Terminal | Security Post | Airplane | Pole | Person |
| **FCN** [87] | 76.40 | 33.61 | 44.91 | 67.89 | 5.36 | 29.76 | 18.27 | **75.06** | 33.65 | **63.34** | 43.70 | 12.82 |
| **PSPNet** [88] | 76.04 | 32.37 | 46.24 | 57.48 | 6.44 | **35.41** | 25.23 | 69.92 | 4.61 | 61.88 | 47.31 | **15.38** |
| **DeepLabV3** [7] | 78.09 | 32.60 | 38.38 | **70.90** | 5.92 | 32.66 | 20.49 | 71.68 | 8.00 | 43.09 | 43.09 | 12.83 |
| **DeepLabV3+** [8] | 78.10 | 32.24 | **50.20** | 55.01 | **8.10** | 22.12 | 24.60 | 70.24 | **46.37** | 58.25 | 49.37 | 5.39 |
| **OCRNet** [9] | 80.52 | 30.92 | 44.73 | 67.68 | 4.76 | 33.61 | 18.47 | 72.17 | 24.00 | 58.03 | 48.74 | 9.34 |
| **HRNet** [10] | **83.29** | **34.47** | 43.55 | 55.72 | 6.39 | 34.37 | **26.34** | 73.72 | 14.62 | 57.98 | **50.26** | 10.40 |
| **BisenetV1** [11] | 60.95 | 22.67 | 45.05 | 53.97 | 3.46 | 0.03 | 15.22 | 38.34 | 0.00 | 21.81 | 30.46 | 0.00 |
| **BisenetV2** [12] | 74.69 | 28.39 | 33.70 | 50.45 | 3.83 | 11.81 | 18.03 | 40.56 | 0.00 | 39.19 | 38.25 | 4.05 |
| **SETR** [13] | 57.96 | 16.17 | 27.80 | 52.87 | 0.10 | 0.00 | 8.89 | 40.63 | 0.00 | 24.09 | 11.99 | 0.00 |
| **Segformer** [14] | 73.24 | 33.22 | 14.93 | 40.21 | 4.84 | 0.16 | 14.51 | 50.86 | 0.00 | 37.96 | 29.19 | 0.00 |
| **Segmentor** [15] | 79.09 | 30.28 | 35.95 | 60.66 | 0.32 | 0.00 | 7.74 | 50.11 | 0.00 | 34.77 | 39.64 | 0.00 |
| **Averaged IoU** | 74.40 | 29.72 | 38.68 | 57.53 | 4.50 | 18.18 | 17.98 | 59.39 | 11.93 | 45.49 | 39.27 | 6.38 |

**Fig. 2.6**  Example images in AGVS-R dataset

R emphasizes capturing the diverse variations of these targets. The challenges in detecting movable objects on the airport ground include multi-scale, target occlusion, view angle changes, illumination changes, etc. Example images from the AGVS-R dataset are shown in Fig. 2.6.

## *2.3.2  Data Collection and Annotation*

**Collection**  The data collection sites are some airports in southwest China. After collecting a large amount of raw data and screening the data following the principle of diversity, the data was found to be somewhat lacking, so some images were collected from the Internet as supplements. The ratio of self-collected data to online data is about 6 to 4. The AGVS-R dataset currently contains 5000 image data, totaling more than 50000 instances for moving targets. We categorize the images in AGVS-R into several subsets according to the challenges. However, each sample contains multiple challenges simultaneously, as the airport is a noncooperative environment with movable targets that are also noncooperative.

**Annotation**  AnyLabeling is used to annotate the images in AGVS-R dataset. The ground truth format of AGVS-R is similar to that of other recognition datasets. But considering that the aircraft is the most important movable target, AGVS-R dataset added three additional labels for aircraft for more granular description. The definition and annotation rules of the three additional labels are described as follows.

The first additional label pertains to the view angle, which represents the aircraft's head orientation in the image. It is defined as the angle between the line from the tail to the head of the aircraft and the positive $X$-axis in the image coordinate system. The view angle of each aircraft may be a value between 0 and 360. For ease of use, we quantize the view angle of the aircraft into eight intervals. Therefore, the value of this label is from 0 to 7.

The second additional label pertains to the occlusion rate. The occlusion rate refers to the proportion of the unpresented parts to the complete area of the aircraft from the same view angle. Similar to the view angle, the occlusion rate should also be a continuous value. For ease of use, we quantify it into three cases: light

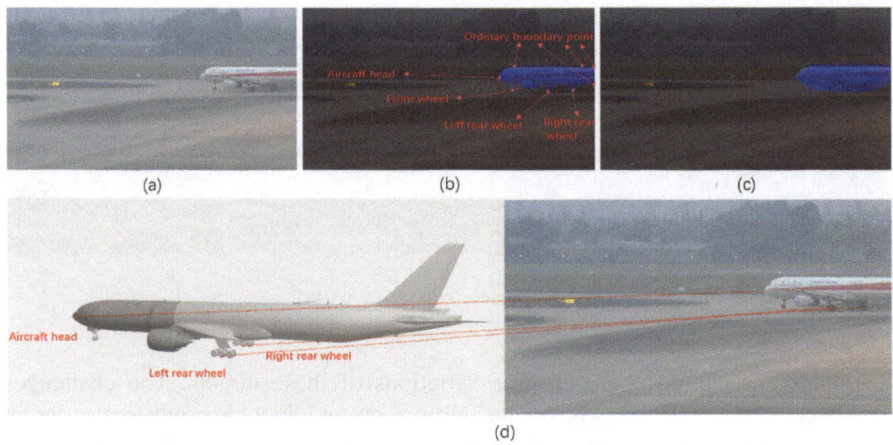

**Fig. 2.7** Annotation process of occlusion degree. (**a**) A frame to be annotated. (**b**) Annotated mask of the aircraft. (**c**) The keypoints of the aircraft. (**d**) The projection of the 3D model based on feature point matching

occlusion, moderate occlusion, and heavy occlusion. Accordingly, the value of this label is from 0 to 2. Note that in order to calculate the occlusion rate, we must outline the object, as we did in the previous two datasets. Furthermore, we must have a reference template that matches the entire shape of the object to calculate the occlusion rate. To this end, we also borrowed the 3D model of the aircraft and a projection method based on matching of aircraft feature points. An accurate segmentation mask is obtained through an annotation method similar to the segmentation dataset, and key feature points are annotated on this basis, as shown in Fig. 2.7b and c. Through keypoint matching, the 3D model can be adjusted and scaled and projected to obtain a pseudo-real mask, as shown in Fig. 2.7d. Finally, the degree of occlusion is calculated according to the next equation,

$$Deg = (S_{Full} - S_{seg})/S_{Full}, \qquad (2.1)$$

where $Deg$ denotes the degree of occlusion, $S_{Full}$ denotes the full area from projection, and $S_{seg}$ denotes the area of the segmented part of the image.

The last additional label is the scale factor. AGVS-R adopts the ratio of the aircraft silhouette coverage area to the total area of the whole image to define the scale of the aircraft. Similarly, the scale factor is also a continuous value, so we also quantify it into several cases for easier processing. Please note that this definition is a relative scale, not the absolute area of the aircraft.

### 2.3.3   Experiments

**Evaluation Metrics** Average precision (AP) is an average of the precision at different recall points, and the larger the AP value indicates that the model is more effective. We use AP and its variants as evaluation metrics on the AGVS-R dataset. For various types of AP indicators, the closer they are to 1, the better the performance.

**Experimental Results** Ten object recognition algorithms were tested on the AGVS-R dataset, including Dynamic RCNN [16], YOLOX-S [17], Centernet [18], Cascade RCNN [19], FCOS [20], DETR [21], Sparse RCNN [22], YOLOF [23], Double-Head RCNN [24], and Mask RCNN [25], and the results are shown in Table 2.3. Table 2.3 also gives the test results on the classic recognition dataset MS-COCO for comparison. It can be seen that the boxAP of each algorithm on the AGVS-R dataset is nearly 20% lower than that on the MS-COCO dataset. On the one hand, it indicates that specific algorithms need to be designed for airport ground movable targets surveillance, and on the other hand, it also shows that the AGVS-R dataset is indeed more challenging than the MS-COCO dataset.

Please note that the AGVS-R dataset, like the AGVS-SS dataset, is still a work in progress, but a preliminary version has been released on the AGVS series website.

**Table 2.3** Performance comparison of 10 state-of-the-art object recognition algorithms on MS-COCO and AVGS-R. Best scores are marking in bold

| Algorithms | Backbone | MS-COCO | AGVS-R | | | | | |
| | | boxAP | boxAP | AP50 | AP75 | APS | APM | APL |
|---|---|---|---|---|---|---|---|---|
| Dynamic RCNN [16] | ResNet-50-FPN | 38.9 | 22.9 | 41.6 | 21.8 | 7.5 | 21.4 | 38.4 |
| YOLOX [17] | Darknet-53 | 40.5 | 20.5 | 37.8 | 19.3 | 5.9 | 17.7 | 36.9 |
| Centernet [18] | ResNet-50-FPN | 40.2 | 23 | 41.2 | 22.2 | 7 | 21 | **39.5** |
| Cascade RCNN [19] | ResNet-50-FPN | 40.4 | 23.8 | 42.3 | **22.9** | 7.9 | **21.5** | 39.2 |
| FCOS [20] | ResNet-50-FPN | 38.5 | 20.6 | 39.5 | 19.1 | 6.1 | 18.6 | 36.4 |
| DETR [21] | DETR | 39.9 | 19.4 | 37.1 | 16.8 | 4 | 14.4 | 36.6 |
| Sparse RCNN [22] | ResNet-50-FPN | **42.8** | **23.9** | 42.5 | **22.9** | **8.8** | 21.4 | 39.4 |
| YOLOF [23] | ResNet-50-C5 | 37.5 | 19.7 | 37.3 | 17.8 | 4.3 | 16.3 | 37.6 |
| Double-Head RCNN [24] | ResNet-50-FPN | 40 | 23.6 | 42.5 | 22.7 | 7.9 | 21.3 | 38.7 |
| Mask RCNN [25] | ResNet-50-FPN | 38.2 | 22.9 | **43.2** | 21.4 | 7.8 | 21.2 | 38.3 |

## 2.4   AGVS-T Dataset

Multiple Object Tracking (MOT) is also an important topic in the field of computer vision. For airports, with the growth of the global economy and population, airports are becoming more and more crowded, and various safety accidents are emerging one after another. Developing intelligent airport applications based on Multiple Aircraft Tracking (MAT), such as conflict warnings, is an effective way to improve the safety level of airports. However, AGVS-T22 [26] points out that most MOT algorithms have varying degrees of performance degradation in airport scenarios. This is mainly because MOT in fundamental research is mostly carried out on pedestrian or vehicle datasets. However, aircraft are unique wingspan targets, and airports have ultra-long monitoring distances, so the MOT algorithms in fundamental research are not suitable for airport scenarios. Therefore, a dedicated airport MOT dataset must be built, and a multi-aircraft tracking algorithm must be designed based on this. In this section, we introduce the tracking dataset in the AGVS series, that is, AGVS-T22 [26]. AGVS-T22 is also constantly improving, and there will be new versions later, such as AGVS-T24.

### 2.4.1   Design and Challenges

As tracking is a spatio-temporal task, the AGVS-T22 dataset consists of video data. The data collection principle aims to capture a wide range of aircraft changes, including various imaging distances and angles, while incorporating different motion modes. The selection criteria emphasize showcasing the diverse challenges encountered in aircraft tracking. Additionally, we paid close attention to video length. Given the relatively slow speed of aircraft in airport scenes, shorter videos may not effectively highlight their movement characteristics. Therefore, the videos in AGVS-T22 generally range from 3 to 5 minutes in length.

Currently, the AGVS-T22 dataset contains 37 long videos, totaling about 120K frames. The videos in AGVS-T22 are named according to the main challenge faced by the video. AGVS-T22 [26] believes that the challenges faced by MAT include three categories: challenges caused by the unique attributes of aircraft, challenges caused by special airport environments, and challenges caused by special imaging modes. These challenges are described as follows.

**Appearance Changes**   The aircraft's design features a distinctive pattern, characterized by its elongated wings and slender fuselage. This results in a significant change in appearance depending on the viewing angle. As depicted in Fig. 2.8, which shows the aircraft during a turn, the continuity of its appearance is disrupted. The multiple long, thin wing protrusions contribute to this visual transformation, making it evident that the aircraft's silhouette alters dramatically as it maneuvers.

**Fig. 2.8** The appearance of an airplane changes significantly when it turns. Adapted with permission from [26]

**Fig. 2.9** Demonstration of the multi-scale problem on the ground. Reprinted with permission from [26]

**Multiple Scales** Due to the vast area of the airport ground, there is an obvious multi-scale problem. In the same field of view, it is possible to see an aircraft with clear fuselage details, or it is possible to see an aircraft that is difficult to distinguish with the naked eye and can only be determined by relying on prior information such as "only aircraft can move on the runway." The scale variation on the airport scene is very large, which is one of the main challenges of MAT. The multi-scale phenomenon on the airport ground is shown in Fig. 2.9.

**Motion Changes** Unlike other ground moving targets such as cars, airplanes have some special motion modes, such as takeoff, landing, and berthing. In addition, due to the huge inertia of the aircraft, there are some special requirements for the movement of the aircraft, such as the targets must maintain a sufficient distance, and there can only be one aircraft on the same runway at a time. Motion modeling is one of the core modules of MOT algorithms, and the existing motion modeling methods may not be applicable to aircraft targets.

**Fig. 2.10** Examples of weather changes and illumination changes in AGVS-T22. Reprinted with permission from [26]

**Weather Changes** Outdoor weather is changeable. Weather conditions such as haze, rain, and cloudiness bring many challenges to aircraft tracking. On haze days, visibility is significantly reduced. On rainy days, water on the ground can cause serious reflections. Under cloudy conditions, there may be significant changes in light and shadow on the imaging plane. Some examples of weather changes are shown in Fig. 2.10.

**Illumination Changes** Airport scenes have a lot of illumination changes. For example, the light intensity varies significantly at different times of the day, but this lighting changes slowly. The smooth surface of the aircraft may cause very bright halos under certain conditions. In addition, due to the huge size of the aircraft, it generally casts a significant shadow under the fuselage. Some examples of illumination changes are also shown in Fig. 2.10.

**Specific Imaging Modes** In addition to visible light imaging, airports utilize other imaging modes, such as infrared and panoramic imaging. This is essential for intelligent airport applications, which require all-weather and panoramic monitoring. These new imaging modes present fresh challenges for MAT. For instance, changes in image properties can lead to a significant drop in the performance of existing MOT algorithms when applied to infrared videos.

### 2.4.2   Data Collection and Annotation

The data collection and screening methods are similar to other datasets. The ground truth format of AGVS-T22 includes the frame number, the target ID number, and the

**Fig. 2.11** Illustration of the annotation process of AGVS-T22. Reprinted with permission from [26]

bounding box coordinates of the target. The annotation process is shown in Fig. 2.11. In order to ensure the accuracy of the annotation, we have established some rules that all annotators must strictly follow. For example, for partially occluded targets, the entire bounding box will be annotated, while for completely occluded objects, the bounding box will not be annotated. There are also some other rules. For example, how small should the target be annotated? Our rule is that if the target outline is discernible to the naked eye, it should be annotated, and otherwise it should not be annotated. Three reviewers are responsible for data review. If two reviewers both think that the data is correctly annotated, it will be sent to the final reviewer for review. Once one of the two reviewers thinks that there is a problem with the annotation, it will be returned and re-annotated.

### 2.4.3  Experiments

**Evaluation Metrics**  In terms of evaluation indicators, we use MOTA as the central evaluation indicator for AGVS-T22. MOTP can reflect the accuracy of detection to some extent. IDF1 can comprehensively reflect the degree of ID change. The formulas for these three indicators are as follows:

$$\text{MOTA} = 1 - \frac{\text{FP} + \text{FN} + \text{IDSW}}{\text{GT}}, \tag{2.2}$$

where GT refers to the number of bounding boxes in the ground truth. It can be seen that MOTA is actually a punishment for FN, FP, IDSW, and then use GT for regular.

$$\text{MOTP} = \frac{\sum_t \sum_i d_i}{C}, \tag{2.3}$$

where $C$ represents the number of matches in frame $t$, and the matching error $d_i$ is calculated for each pair of matches and represents the distance between the target $O_i$ and the assumed pairing position in frame $t$.

$$\text{IDF1} = \frac{2 \times \text{IDTP}}{2 \times \text{IDTP} + \text{IDFP} + \text{IDFN}}, \tag{2.4}$$

**Table 2.4** Performance comparison of eight tracking algorithms on MOT17 and AGVS-T22. Best scores are making in bold. Adapted with permission from [26]

| Methods | MOT17 | | | AGVS-T22 (proposed) | | |
|---|---|---|---|---|---|---|
| | MOTA ↑ | MOTP ↑ | IDF1 ↑ | MOTA ↑ | MOTP ↑ | IDF1 ↑ |
| SORT [27] | 59.8 | 69.5 | 53.8 | 40.3 | **35.5** | 47.3 |
| DeepSORT [28] | 61.4 | 63.5 | 62.5 | 33.6 | 28.1 | 38.2 |
| OC-SORT [29] | 80.3 | **79.8** | **76.3** | 64.0 | 26.2 | 56.1 |
| ByteTrack [30] | 76.8 | 69.0 | 71.6 | **65.0** | 24.3 | 35.9 |
| BoT-SORT [31] | **80.6** | 77.2 | 76.1 | 64.8 | 26.5 | **56.2** |
| JDE [32] | 64.6 | 60.3 | 55.8 | 53.1 | 22.3 | 38.4 |
| FairMOT [33] | 60.4 | 60.4 | 72.3 | 53.9 | 31.3 | 35.0 |
| CenterTrack [34] | 76.3 | 49.5 | 60.0 | 60.9 | 13.5 | 34.3 |

$$IDP = IDTP/(IDTP + IDFP), \qquad (2.5)$$

$$IDR = IDTP/(IDTP + IDFN), \qquad (2.6)$$

where IDTP (True Positive) is the number of correctly matched identities, IDFP (False Positive) is the number of incorrectly matched identities, and IDFN (False Negative) is the number of missed identities.

**Experimental Results**　　Eight state-of-the-art MOT algorithms were tested on the AGVS-T22 dataset, and the results are shown in Table 2.4. The eight algorithms are SORT [27], DeepSORT [28], OC-SORT [29], ByteTrack [30], BoT-SORT [31], JDE [32], FairMOT [33], and CenterTrack [34]. The MOT17 dataset is also tested for comparison. From this table, we can clearly see that almost all algorithms perform better than AGVS-T22 on MOT17. This once again shows that it is necessary to build a dedicated airport tracking dataset and conduct research on MAT.

## 2.5　AGVS-AR Dataset

Action recognition is a crucial area of computer vision. Automatically detecting and recognizing aircraft actions can enhance the operational efficiency and safety of airports. For instance, real-time monitoring of aircraft behavior can help prevent potential runway incursions and conflicts. Categories of aircraft actions at airports include takeoff, landing, taxiing, docking, undocking, and waiting. While there are some studies on video action recognition, most focus on general scenarios or specific action types, with limited research on airport aircraft actions. Furthermore, existing datasets primarily emphasize the recognition of people and vehicles, leaving a gap for datasets specifically tailored to airport aircraft actions. This shortage restricts research depth in this area and hinders the development of related technologies.

In order to promote the development of aircraft action recognition technology, we propose a new dataset AGVS-AR, which contains video sequences of various actions of different aircraft at airports under different weather conditions, different airport scenes, different aircraft scales, and different aircraft. This dataset mainly emphasizes three aspects. First, it emphasizes the unique properties of aircraft and airport environments. Unlike general datasets that focus on nonrigid targets such as people, aircraft are rigid targets, and many of their actions need to be analyzed not only individually but also in combination with the surrounding environment. For example, if the aircraft is turning or going straight, it is taking off in the runway area, while it may be docking in the apron area. Secondly, it covers various actions that may occur during the taxiing process, including meta-actions such as straight line and turning, as well as advanced actions such as docking, undocking, and takeoff. Finally, each video in this dataset features multiple aircraft or various aircraft action patterns. Thus, it resembles an MOT dataset that necessitates simultaneous detection and data association. When using this dataset, it is essential to concurrently locate and classify the aircraft's actions.

## 2.5.1 Design and Challenges

Similar to other datasets, the design principle of AGVS-AR is to cover as many aircraft action categories as possible and include various challenging factors for aircraft action recognition. As you can see, dataset design and challenges are two sides of the same problem. Based on long-term observation of airport scenes, we summarized the dataset design scheme and the various challenges faced by aircraft action recognition, which are described as follows.

**Multi-level Actions** In AGVS-AR, we divide the aircraft's actions into two different levels, meta-actions and advanced actions. Meta-actions refer to the general basic actions of the aircraft, including straight, turning, flying, and stationary. Advanced actions refer to specific actions with certain behavioral significance, including docking, undocking, takeoff, landing, queuing, loading, unloading, bridge docking, bridge disconnection, tractor docking, tractor disconnection, etc. It can be seen that meta-actions are the basis of advanced actions, and different permutations and combinations of meta-actions constitute advanced actions. It is impossible to judge the current behavior status of the aircraft only from meta-actions.

**Multiple Targets and Multiple Actions** There may be multiple targets in a video, and each target may also have multiple actions at the same time. For example, as shown in Fig. 2.12, the video represented in the left picture contains multiple aircraft, and each aircraft has different actions, while the video shown in the right picture contains multiple advanced actions on a single aircraft body, simultaneously unloading passengers and unloading cargoes. This multi-target, especially simultaneous multi-behavior phenomenon, poses a huge challenge to action recognition.

**Fig. 2.12**  The left picture shows a video containing multiple aircraft with a single behavior, while the right picture shows a video containing only one aircraft but with two different actions

**Fig. 2.13**  Four examples in AGVS-AR with varying degrees of action ambiguity

**Action Ambiguity**  The samples in AGVS-AR are not all complete, and some may be ambiguous. We believe that an excellent action recognition algorithm should be robust to this problem. Some examples of ambiguous action are shown in Fig. 2.13. In Fig. 2.13a, the aircraft in the circle is flying from the right side of the runway to the left side, and its action is most likely to be landing, but the context information provided by the video is insufficient, and the possibility of takeoff cannot be ruled out. In Fig. 2.13b, although the two aircraft in the circle are missing most of the main body of the aircraft, according to experience, their most likely actions are docking and unloading. In Fig. 2.13c, there are people in the circle boarding the shuttle bus, and according to experience, they are unloading, but the main body of the aircraft is missing. In Fig. 2.13d, the aircraft in the circle is taxiing toward the terminal, and its action may be a berth. An excellent behavior recognition algorithm should be able to make reasonable inferences about the above ambiguous samples.

**Fig. 2.14** Some airport scenes with complex lighting and weather in AGVS-AR

**Appearance Changes**  The unique appearance and appearance changes of aircraft are also an important challenge. For example, the appearance of different aircraft is mostly white, and the size and shape are similar, which makes it difficult to identify each aircraft. In addition, the imaging scale of aircraft may vary greatly, and even multiple scales may exist in the same scene at the same time. A certain degree of appearance change can be seen in each example in the previous part of this book.

**Environmental Factors**  Environmental factors such as weather changes and illumination changes are also major challenges for action recognition. Common weather conditions include sunny days, rainy days, cloudy days, and haze days. Under haze conditions, due to the vast area of the airport, the target resolution is usually low, and as the distance increases, the air transparency decreases significantly, resulting in severe video blur. On rainy days, water on the ground can cause severe reflections. Due to the smooth surface of the fuselage, obvious reflections can be caused at certain angles, which distorts the appearance of the target. Weather and illumination changes have been described in detail in the previous part of this book. Some other examples of such changes are given in Fig. 2.14.

The AGVS-AR dataset is in production, and we will soon release a preliminary version of AGVS-AR on the AGVS series website.

## 2.6  FGVC-Aircraft Dataset

The FGVC-Aircraft dataset [35] is a dataset specifically designed for Fine-Grained Visual Classification of Aircraft (FGVC-Aircraft). Although the samples in the

FGVC-Aircraft dataset are all aircraft, they are not all collected in the airport area. Furthermore, almost all aircraft are in the center of the image at a close distance and from the side, with almost no distance and angle changes, and almost no weather changes and occlusion. Therefore, this dataset is mainly used for aircraft image classification research and is not suitable for airport ground video surveillance. However, considering that the existing airport dataset is very limited, we will also give a brief introduction to it here. The FGVC-Aircraft dataset contains about 10,000 images, covering 102 different types of aircraft, including civil and military aircraft. Each image is accompanied by detailed label information, such as aircraft type and part location. Due to the similarity in appearance of aircraft, this makes the FGVC-Aircraft dataset an ideal choice for training and evaluating classification models, especially for classification tasks of objects with similar appearance.

## 2.7   ALERT Dataset

ALERT [36] is a person reidentification dataset. Person reidentification aims to identify the same pedestrian captured under different camera perspectives or time periods. Person reidentification is widely used in video surveillance, intelligent transportation, and other fields. Due to the complex interaction between pedestrians and the environment, such as posture changes, occlusion, and lighting differences, pedestrian reidentification faces many challenges. In fact, pedestrian reidentification has little to do with airport ground video surveillance, because the main body of airport ground video surveillance is aircraft, not pedestrians. However, because the samples in ALERT are all collected at the airport terminal, and both aircraft tracking and pedestrian tracking face the problem that different instances are too similar, we also give a brief introduction here.

The ALERT dataset uses six cameras to collect data. These cameras cover the central security checkpoint area and three concourses of the terminal. Each camera has a resolution of $768 \times 432$ pixels and shoots video at 30 frames per second. Each camera recorded up to 12 hours of video from 8 am to 8 pm. Each long video was randomly split into 45-minute-long video clips. In total, tracks corresponding to 9651 people were extracted from all short video clips. The number of bounding boxes in the dataset is 39902, with an average of about 3.13 images per person. The sizes of the detected bounding boxes range from $130 \times 54$ to $403 \times 166$. Among the 9651 people, 1382 people are paired on at least two cameras.

## References

1. Zhang X, Shu C, Li S, Wu C, Liu Z (2022) AGVS: A new change detection dataset for airport ground video surveillance. IEEE Trans Intell Transp Syst 23(11):20588–20600. https://doi.org/10.1109/TITS.2022.3184978

2. St-Charles PL, Bilodeau GA, Bergevin R (2015) SuBSENSE: A universal change detection method with local adaptive sensitivity. IEEE Trans Image Process 24(1):359–373. https://doi.org/10.1109/TIP.2014.2378053

3. Barnich O, Van Droogenbroeck M (2009) ViBE: A powerful random technique to estimate the background in video sequences. In: 2009 IEEE International Conference on Acoustics, Speech and Signal Processing (ICASSP). IEEE, Taipei, Taiwan, pp 945–948

4. Kim K, Chalidabhongse TH, Harwood D, Davis L (2005) Real-time foreground–background segmentation using codebook model. Real-Time Imaging 11(3):172–185. https://doi.org/10.1016/j.rti.2004.12.004

5. Wang Y, Luo Z, Jodoin P-M (2017) Interactive deep learning method for segmenting moving objects. Pattern Recogn Lett 96: 66–75. https://doi.org/10.1016/j.patrec.2016.09.014

6. Lim LA, Keles HY (2018) Foreground segmentation using a triplet convolutional neural network for multiscale feature encoding. Pattern Recogn Lett 112:256–262. https://doi.org/10.1016/j.patrec.2018.08.002

7. Chen L-C, Papandreou G, Schroff F, Adam H (2017) Rethinking atrous convolution for semantic image segmentation. arXiv:1706.05587, 2017

8. Chen L, Zhu Y, Papandreou G, Schroff F, Adam H (2018) Encoder-decoder with atrous separable convolution for semantic image segmentation. In: 2018 European Conference on Computer Vision (ECCV). Springer, Munich, pp 833–851

9. Yuan Y, Chen X, Chen X, Wang J (2021) Segmentation transformer: object-contextual representations for semantic segmentation. arXiv:1909.11065, 2019

10. Wang J, Sun K, Cheng T (2020) Deep high-resolution representation learning for visual recognition. IEEE Trans Pattern Anal Mach Intell 43(10):3349–3364. https://doi.org/10.1109/TPAMI.2020.2983686

11. Yu C, Gao C, Wang J (2021) Bisenet v2: bilateral network with guided aggregation for real-time semantic segmentation. Int J Comput Vis 129:3051–3068. https://doi.org/10.1007/s11263-021-01515-2

12. Yu C, Wang J, Peng C, Gao C, Yu G, Sang N (2018) BiSeNet: bilateral segmentation network for real-time semantic segmentation. In: 2018 European Conference on Computer Vision (ECCV). Springer, Munich, pp 334–349

13. Zheng S, Lu J, Zhao H, Zhu X, Luo Z, Wang Y, Fu Y, Feng J, Xiang T, Torr PHS, Zhang L (2021) Rethinking semantic segmentation from a sequence-to-sequence perspective with transformers. In: 2021 IEEE/CVF Conference on Computer Vision and Pattern Recognition (CVPR). IEEE, Nashville, Tennessee, pp 6877–6886

14. Xie E, Wang W, Yu Z (2021) SegFormer: simple and efficient design for semantic segmentation with transformers. In: 2021 Advances in Neural Information Processing Systems (NIPS), vol 34. MIT Press, New York, pp 12077–12090

15. Strudel R, Garcia R, Laptev I, Schmid C (2021) Segmenter: transformer for semantic segmentation. In: 2021 IEEE/CVF International Conference on Computer Vision (ICCV). IEEE, Montreal, Quebec City, pp 7242–7252

16. Zhang H, Chang H, Ma B, Wang N, Chen X (2020) Dynamic R-CNN: towards high quality object detection via dynamic training. In: 2020 European Conference on Computer Vision (ECCV). Springer, Berlin, pp 260–275

17. Ge Z, Liu S, Wang F, Li Z, Sun J (2021) YOLOX: Exceeding YOLO Series in 2021. arXiv:2107.08430, 2021.

18. Duan K, Bai S, Xie L, Qi H, Huang Q, Tian Q (2019) CenterNet: keypoint triplets for object detection. In: 2019 IEEE/CVF International Conference on Computer Vision (ICCV). IEEE, Seoul, Korea (South), pp 6568–6577

19. Cai Z, Vasconcelos N (2018) Cascade R-CNN: delving into high quality object detection. In: 2018 IEEE/CVF Conference on Computer Vision and Pattern Recognition (CVPR). IEEE, Salt Lake City, UT, pp 6154–6162

20. Tian Z, Shen C, Chen H, He T (2020) FCOS: a simple and strong anchor-free object detector. IEEE Trans Pattern Anal Mach Intell 44(4):1922–1933. https://doi.org/10.1109/TPAMI.2020.3032166

21. Carion N, Massa F, Synnaeve G, Usunier N, Kirillov A, Zagoruyko S (2020) End-to-End object detection with transformers. In: 2020 European Conference on Computer Vision (ECCV). Springer, Berlin, pp 213–229
22. Sun P, Zhang R, Jiang Y, Kong T, Xu C, Zhan W, Tomizuka M, Li L, Yuan Z, Wang C, Luo P (2021) Sparse R-CNN: end-to-end object detection with learnable proposals. In: 2021 IEEE/CVF Conference on Computer Vision and Pattern Recognition (CVPR). IEEE, Nashville, TN, pp 14449–14458
23. Chen Q, Wang Y, Yang T, Zhang X, Cheng J, Sun J (2021) You only look one-level feature. In: 2021 IEEE/CVF Conference on Computer Vision and Pattern Recognition (CVPR). IEEE, Nashville, Tennessee, pp 13034–13043
24. Wu Y, Chen Y, Yuan L, Liu Z, Wang L, Li H, Fu Y (2020) Rethinking classification and localization for object detection. In: 2020 IEEE/CVF Conference on Computer Vision and Pattern Recognition (CVPR). IEEE, Seattle, Washington, pp 10183–10192
25. He K, Gkioxari G, Dollár P, Girshick R (2017) Mask R-CNN. In: 2017 IEEE/CVF International Conference on Computer Vision (ICCV). IEEE, Venice, pp 2961–2969
26. Li T, Zhang X, Tang Z, Liu Y (2023) AGVS-T22: A new multiple object tracking dataset for airport ground video surveillance. In: 2023 IEEE 26th International Conference on Intelligent Transportation Systems (ITSC). IEEE, Bilbao, pp 4380–4387
27. Bewley A, Ge Z, Ott L, Ramos F, Upcroft B (2016) Simple online and realtime tracking. In: 2016 IEEE International Conference on Image Processing (ICIP). IEEE, Phoenix, Arizona, pp 3464–3468
28. Wojke N, Bewley A, Paulus D (2017) Simple online and realtime tracking with a deep association metric. In: IEEE International Conference on Image Processing(ICIP). IEEE, Beijing, pp 3645–3649
29. Cao J, Pang J, Weng X, Khirodkar R, Kitani K (2023) Observation-Centric SORT: Rethinking SORT for robust multi-object tracking. In: IEEE/CVF Conference on Computer Vision and Pattern Recognition (CVPR). IEEE, Vancouver, pp 9686–9696
30. Zhang Y, Sun P, Jiang Y, Yu D, Weng F, Yuan Z, Luo P, Liu W, Wang X (2022) ByteTrack: multi-object tracking by associating every detection box. In: 2022 European Conference on Computer Vision (ECCV). Springer, Cham. Lecture Notes in Computer Science, vol 13682, pp 1–21
31. Aharon N, Orfaig R, Bobrovsky B-Z (2022) BoT-SORT: robust associations multi-pedestrian tracking. arXiv:2206.14651, 2022.
32. Wang Z, Zheng L, Liu Y, Li Y, Wang S (2020) Towards real-time multi-object tracking. In: 2020 European Conference on Computer Vision (ECCV). Springer, Cham, pp 107–122
33. Zhang Y, Wang C, Wang X, Zeng W, Liu W (2021) FairMOT: On the fairness of detection and re-identification in multiple object tracking. Int J Comput Vis 129(11):3069–3087. https://doi.org/10.1007/s11263-021-01513-4
34. Zhou X, Koltun V, Krahenbuhl P (2020) Tracking objects as points. In: 2020 European Conference on Computer Vision (ECCV). Springer, Cham, pp 474–490
35. Maji S, Rahtu E, Kannala J, Blaschko M, Vedaldi A (2013) Fine-grained visual classification of aircraft. arXiv:1306.5151, 2013
36. Karanam S, Gou M, Wu Z, Rates-Borras A, Camps O, Radke RJ (2019) A systematic evaluation and benchmark for person re-identification: features, metrics, and datasets. IEEE Trans Pattern Anal Mach Intell 41(3):523–536. https://doi.org/10.1109/TPAMI.2018.2807450

# Chapter 3
# Algorithms

**Abstract** Building on the AGVS dataset series from the previous chapter, our team conducted initial research on computer vision algorithms for airport ground video surveillance. We aimed to leverage unique prior information from airport environments to enhance algorithm performance. This chapter outlines the design principles for airport-specific computer vision algorithms and presents our findings in three key areas: segmentation, recognition, and tracking in airport ground video surveillance.

## 3.1 Principles

In airports, there are not only image and video data, but also data from other data sources, such as radar point cloud data from Surface Surveillance Radar (SSR), satellite positioning data from aircraft transponders, 3D model data of various aircraft, electronic map data, etc. These heterogeneous data can provide some information that images and videos do not have. For example, radar point cloud data contains the 3D coordinate information of the target, while satellite positioning data contains the ID information of the target. In addition, electronic maps also depict terrain such as runways and taxiways and their structural relationships. Because of the complementarity of multisource data, we believe that the effective use of this nonimage information can improve the performance of airport computer vision algorithms.

We propose design principles for computer vision algorithms for airport ground video surveillance. That is, using various airport prior information to guide the development of computer vision algorithms. For example, 3D digital models for moving objects can be used to guide airport semantic segmentation. Such algorithms are particularly effective for intelligent airport applications because they utilize airport-specific data and can often significantly improve performance. However, such algorithms are tailored for airports and therefore lack generalizability and cannot be directly applied to scenarios outside of airports. Nevertheless, the algorithm design principles we propose are still applicable to other scenarios. For example, in maritime traffic scenarios, unique environmental factors such as

31
X. Zhang et al., *Airport Ground Video Surveillance*,
https://doi.org/10.1007/978-981-96-2310-5_3

water surface reflection patterns can also serve as prior information for designing computer vision algorithms for maritime traffic.

Next, we will discuss several representative examples of airport-specific prior information and explore potential applications of this data.

### 3.1.1   3D Digital Model Prior

In addition to image and video data, there are other data related to the targets in the scene, such as 3D digital models. When the scene is limited and the target type is known, we can obtain the 3D digital model of the target through various means. These digital models can provide very useful prior information for computer vision tasks. Take aircraft segmentation as an example. Since the aircraft is a typical wingspan target, there are often serious segmentation defects in the wingspan area, such as the side wings and tail parts of the aircraft. However, the 3D digital model of the aircraft can provide complete shape prior information without any defects, so using the 3D digital model to constrain the aircraft segmentation helps to improve the segmentation accuracy. One idea of using this prior information is to project the 3D model onto multiple 2D planes to generate a series of 2D images. Then, various similarity metrics are used to identify the projection that best matches the shape of the aircraft in the image, and use it as shape prior knowledge in methods such as graph neural networks or level sets, as shown in Fig. 3.1.

### 3.1.2   Interactive Shape Prior

The previously discussed 3D digital model serves as a shape prior. We now introduce a more flexible shape prior that relies on manual interaction. In the airport scene, certain semantic objects have unique shapes that can be strongly constrained by simple user input. For example, the general shape of an aircraft can be depicted by drawing two intersecting lines connecting the nose, tail, and wing, as shown in Fig. 3.2. Although this cross-shape constraint is not as strong as the 3D model, it can effectively capture the most challenging wingspan area of the aircraft. Moreover,

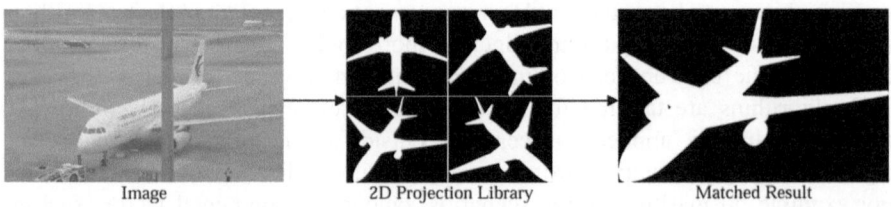

Image                           2D Projection Library                    Matched Result

**Fig. 3.1**  Illustration of 3D digital model and the matching process based on projection

**Fig. 3.2**  Illustration of interactive shape prior

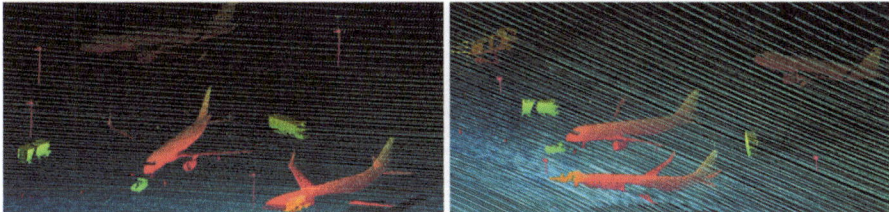

**Fig. 3.3**  Visualization of radar point cloud data from different angles

this annotation method is very intuitive and easy to operate. We can use such interactive shape prior knowledge to guide computer vision tasks. For example, a conditional diffusion model can be created based on the cross-constraint to generate semantic segmentation from random Gaussian noise.

### 3.1.3  Radar Data Prior

Many airports utilize ground surveillance radars to generate radar point cloud data. Unlike 2D image data, radar point clouds offer additional 3D coordinate information, as shown in Fig. 3.3. Radar imaging is reliable, performing well in low light and adverse weather. However, point cloud data is often sparse, while image data is high resolution and detailed. Thus, integrating radar point cloud data with image data is advantageous, as they complement each other and improve airport computer vision tasks.

We use semantic segmentation to illustrate the integration of radar and image data. Since point cloud and image data are heterogeneous, data registration is necessary. To incorporate point cloud data into image semantic segmentation frameworks, it must first be converted into a compatible format. A depth map, which is a two-dimensional image providing distance information between objects and the imaging device, can be easily generated from point cloud data via projection transformation, as shown in Fig. 3.4. The depth map, formatted like an ordinary image and containing target appearance information, allows for effective fusion of the two

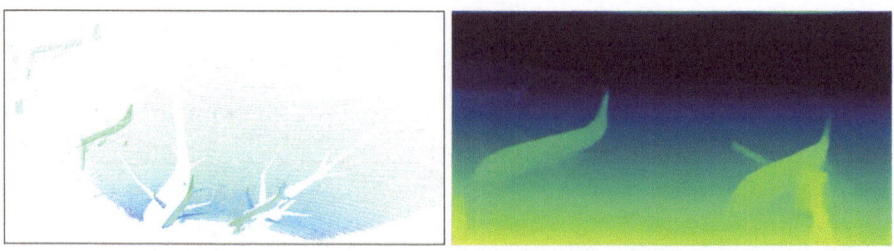

**Fig. 3.4** Depth map obtained by projecting point cloud data to 2D plane

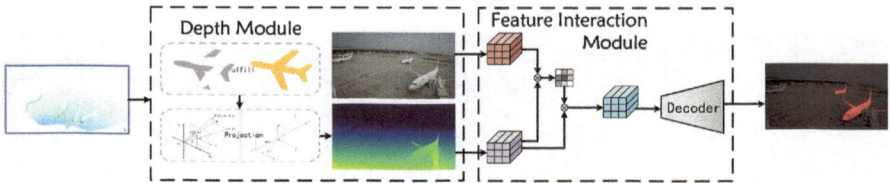

**Fig. 3.5** Flowchart of an idea for semantic segmentation based on "radar + image"

data types through feature interaction, resulting in improved semantic segmentation outcomes. An example of this fusion is illustrated in Fig. 3.5, comprising a depth module and a feature interaction module.

In the depth module, point cloud completion methods first fill in missing areas, followed by converting the point cloud data into depth data using a compression projection algorithm. In the feature interaction module, cross-attention aligns and interacts with point cloud and image features, enhancing feature descriptiveness through contrastive learning. Finally, the fused features are decoded to produce semantic segmentation results.

### 3.1.4   Satellite Positioning Prior

The airport also has satellite positioning information. For example, the ADS-B data broadcast in real time by the transponder installed on the head of the aircraft contains the satellite positioning information of the aircraft. If the satellite positioning data in the ADS-B data is projected onto the image plane, the ADS-B data can be converted into an image trajectory, as shown in Fig. 3.6a. A significant advantage of satellite positioning data is that it is not affected by observation distance and weather conditions. Satellite signals can be received regardless of the target's location or weather conditions.

Furthermore, the satellite trajectory conveys scale information about the aircraft, illustrated in Fig. 3.6b. According to perspective imaging principles, objects appear larger when closer and smaller when farther away. Thus, distant aircraft have a

**Fig. 3.6**  Illustration of satellite trajectory projected from ADS-B data onto the image plane

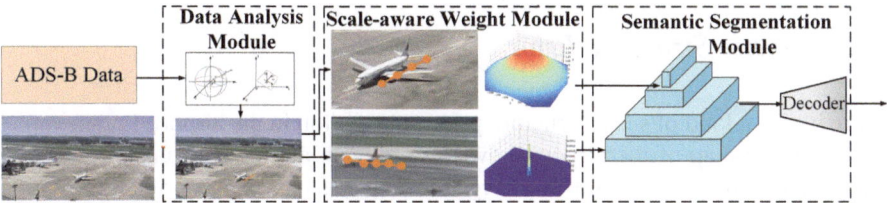

**Fig. 3.7**  Flowchart of an idea for airport semantic segmentation based on satellite positioning information

smaller imaging scale, while those nearby appear larger. Since aircraft are large and typically glide at a constant speed, the sparsity of the satellite positioning trajectory correlates with the target's imaging scale. Larger aircraft exhibit a relatively sparse satellite trajectory, while smaller ones show a denser trajectory, as depicted in Fig. 3.6b. Therefore, the satellite positioning trajectory can also provide scale prior information, which is beneficial for various computer vision tasks, particularly for small-scale targets. For instance, we propose an idea for airport semantic segmentation based on satellite positioning data, as demonstrated in Fig. 3.7.

This idea consists of three modules: data parsing module, scale weighting module, and semantic segmentation module. The data parsing module extracts satellite positioning data and projects it onto the image plane to generate motion trajectories. The scale weighting module is the core module, which creates a scale weighting function for each target identified by the parsing module. This function assigns smaller weights to targets with larger scales (easier to segment) and larger weights to targets with smaller scales (more difficult to segment). Assigning larger weights to small targets helps improve the segmentation accuracy of small targets. The scale weighting function is defined as a two-dimensional Gaussian function, and the weight and coverage area are adjusted according to the scale factor. Finally, the scale weighting function is used in the semantic segmentation module to enhance the features of small targets on the feature pyramid, and the final segmentation result is obtained after decoding.

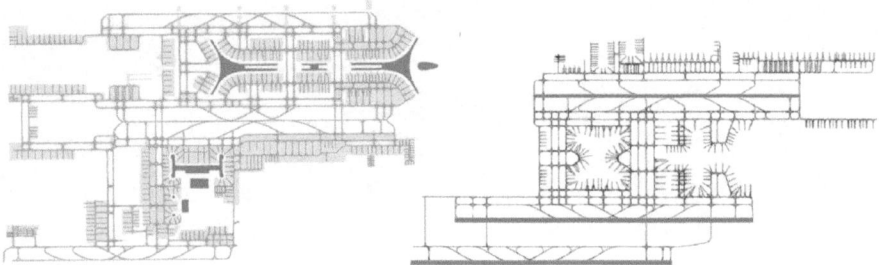

**Fig. 3.8**  An example of airport electronic map

### 3.1.5  Scene Structure Prior

There is also a kind of scene structure prior information in the airport. For example, the electronic map of the airport shows terrain such as runways, taxiways, grass, aprons, and terminals, as shown in Fig. 3.8. The scene structure prior information is closely related to the active targets in the scene. For example, airplanes can only appear in certain specific areas, such as aprons and runway areas, but not in other areas, such as lawns and terminal areas. In addition, airplanes exhibit different movement patterns in different areas. They can only move in a straight line on the runway, may turn at intersections, and finally stop next to the apron bridge. Therefore, structural prior information provides prior constraints on the position and movement of active targets, which is beneficial for various computer vision tasks.

Above we have summarized several prior knowledge related to airport scenes. Airport scenes also include other prior knowledge, such as color prior knowledge. Next, we will introduce our team's results in segmentation, recognition, and tracking of airport ground video surveillance. Most methods are driven by scene prior knowledge. In our "scene prior driven" research, we are still in the early stages and use only limited prior information. Due to the limitations of the data source, some prior information, such as radar data prior, has not been fully explored.

## 3.2  Segmentation

### 3.2.1  Related Work

Segmentation is one of the fundamental topics in computer vision. Segmentation includes multiple sub-directions, from the initial binary change detection, to video object segmentation, to the recent multi-classification semantic segmentation, etc. Here, we only give a very brief introduction to the development of semantic segmentation. The task of semantic segmentation is to assign semantic labels to each pixel in an image. It has a wide range of applications, including autonomous driving, video surveillance, medical image analysis, and augmented reality. The development of semantic segmentation has evolved from traditional methods to modern deep learning-based methods. Semantic segmentation methods can be roughly divided into three categories based on input data and training methods: Supervised Image Semantic Segmentation (SISS), Imprecise Supervised Image Semantic Segmentation (ISISS), and Video Semantic Segmentation (VSS).

SISS [1–5] relies on fully annotated pixel-level labels for training, making it the most accurate but also the most resource-intensive method. Traditional SISS faces challenges such as manually designed features, which often limit the expression of complex semantic information. With the advent of deep learning, SISS has made significant progress, mainly through the use of Fully Convolutional Networks (FCNs) [1], which replace fully connected layers with convolutional layers, enabling the network to process images of any size while preserving spatial information. FCNs laid the foundation for modern SISS methods by enabling end-to-end pixel-level classification.

Imprecisely Supervised Image Semantic Segmentation (ISISS) emerged as a response to the high cost and effort of creating fully annotated datasets for SISS. It leverages training samples with less precise annotations, such as image-level labels [6–9] bounding boxes [10, 11], or even completely unsupervised data [12]. Weakly supervised techniques rely on image-level annotations typically involve a two-stage process: First, a classification network is trained with image-level labels, and then pseudosegmentation masks are generated from Class Activation Maps (CAMs) [9]. Another approach in ISISS is semi-supervised learning [13–15], which combines a small set of labeled data with a large set of unlabeled data to improve model performance. This is often achieved through techniques such as pseudo-labeling and consistency regularization. ISISS is particularly useful in cases where accurate annotations are difficult or expensive to obtain.

Video Semantic Segmentation (VSS) [16–20] extends the semantic segmentation task to video data, incorporating temporal information to improve accuracy and efficiency. Early VSS methods processed each frame independently, but more recent methods exploit temporal consistency between frames. Methods such as Clockwork FCN [19] and Deep Feature Flow (DFF) [20] have been developed to address the unique challenges of video data. For example, DFF uses optical flow to propagate features from keyframes to subsequent frames, significantly reducing computational overhead without sacrificing accuracy. More advanced VSS techniques include the use of spatio-temporal networks, which integrate spatial and temporal features to provide more accurate segmentation results. These networks typically employ Long Short-Term Memory (LSTM) modules or 3D convolutions to model temporal dependencies across frames.

### *3.2.2   Extended Motion Diffusion-Based Change Detection*

Our group proposed an extended motion diffusion-based change detection method
for airport ground video surveillance [21]. Change detection is a classic segmentation problem, whose goal is to distinguish moving objects from static backgrounds
in a video. Therefore, change detection is a typical binary classification problem.
Our experiments show that the performance of change detection in airports is very
poor. As shown in Fig. 3.9, there is a serious detection defect in airport change
detection. We noticed that there is a structural prior information in airports, that is,
airplanes can only travel along straight roads such as runways or taxiways, and the
various straight roads in airports are either parallel or perpendicular to each other.
This is clearly visible from the satellite image of the airport from a bird's-eye view,
as shown in Fig. 3.10.

We consider using the structural prior information shown in Fig. 3.10 to improve
the performance of airport change detection. The basic idea of the paper [21] is to
generate and diffuse samples based on structural prior information. Because of the
constraints of structural prior information, such samples can effectively compensate
for the detection defects in Fig. 3.9. The process and experimental results of method
[21] are described as follows.

**Approach**   The method [21] includes two important modules, the structural information extraction module and the structural information application module. We
noticed that due to perspective imaging, the straight roads in the airport are not

**Fig. 3.9** Demonstration of change detection in airport ground. Reprinted with permission from
[21]

**Fig. 3.10** The satellite image of the airport ground. Reprinted with permission from [21]

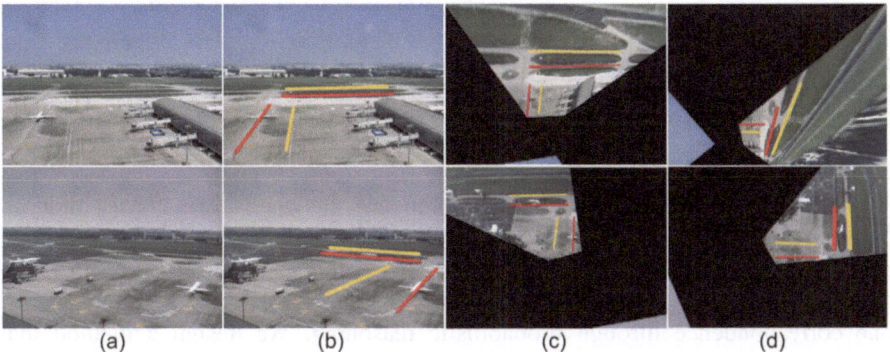

(a)        (b)        (c)        (d)

**Fig. 3.11** (**a**) Two frames from the AGVS-S dataset. (**b**) Manually marked straight runways and taxiways. (**c**) and (**d**) Transformed images from (**b**) to a top view. Reprinted with permission from [21]

actually parallel or perpendicular to each other on the imaging plane, as shown in Fig. 3.11b. Therefore, in the structural information extraction module, we designed a method based on homography transformation to convert the side-view airport image into a top-view perspective, so as to achieve the effect of parallel or perpendicular straight roads, as shown in Fig. 3.11c and d. Homography transformation can describe the transformation relationship between planes. Because the airport ground is a very standard plane in both the top-view and side-view perspectives, the use of homography transformation can well achieve our goal.

Due to the huge mass and inertia of the aircraft, the aircraft must keep consistent with the direction of the road during taxiing; otherwise, it may cause serious safety accidents. In the structural information extraction module, we already know the direction of the straight road in the airport, which actually knows the direction of the aircraft's movement. In other words, based on the airport structure prior information, we can predict the aircraft's motion vector. In the structural information application module, we generate and diffuse samples based on the aircraft's motion vector in order to compensate for detection defects. The principle of the structural information application is shown in Fig. 3.12.

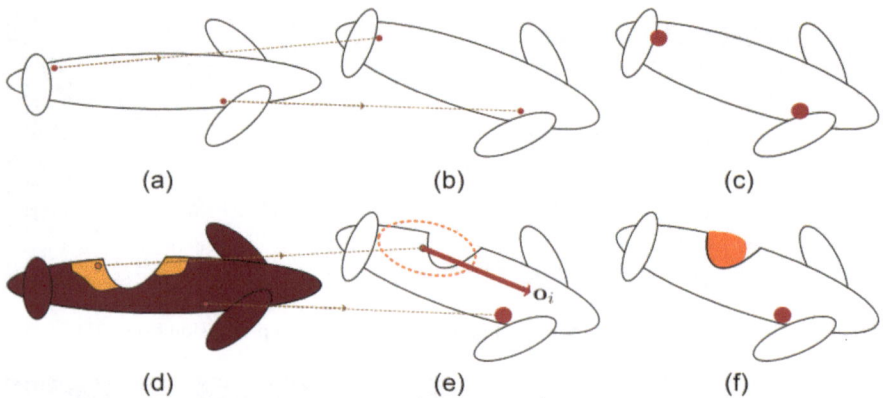

**Fig. 3.12** The principles of [21]. One-to-one correspondence from (**a**) to (**b**) based on optical flow. (**c**) The one-to-many correspondence in motion diffusion. (**d**) Classification of detected pixels. (**e**) Sample synthesis. (**f**) Sample screening. Reprinted with permission from [21]

Based on optical flow tracking, we can get the one-to-one correspondence between the pixels in Fig. 3.12a and b. Considering that optical flow is usually not accurate enough, motion diffusion assumes that each pixel corresponds to a region, as shown in Fig. 3.12c, and then obtains a relatively accurate one-to-one correspondence through probabilistic reasoning. We design a method that exploits the one-to-many correspondence in motion diffusion. If a pixel is close to a detection defect region, its one-to-many corresponding region may cover the detection defect. In this case, the new samples synthesized in the corresponding region may compensate for the detection defect. Specifically, we first divide the detected pixels into two categories, as shown in Fig. 3.12d. For pixels close to the detection defect, a relatively large elliptical area is predicted, as shown in Fig. 3.12e, and new samples are synthesized in this area. Next, the synthesized samples are screened to keep only the truly useful samples, as shown in Fig. 3.12f. In particular, the main axis of the elliptical area in Fig. 3.12e is parallel to the direction vector of the aircraft, which is more conducive to the generated samples covering the detection defect area. The direction vector of the aircraft is obtained in the previous structural information extraction module.

**Experimental Results** A specific example of the above method is given in Fig. 3.13. Figure 3.13a shows the result of the initial segmentation. Figure 3.13b shows the result of the optical flow tracking. Figure 3.13c shows the direction vector of the aircraft and the motion diffusion area. Figure 3.13d and e demonstrates the synthesized samples and the filtered samples. Figure 3.13f shows the segmentation result of the model trained based on the new samples. It can be seen that the new model can completely segment the aircraft, but the new model cannot handle the shadow area under the fuselage. Some other experimental results are shown in Fig. 3.14.

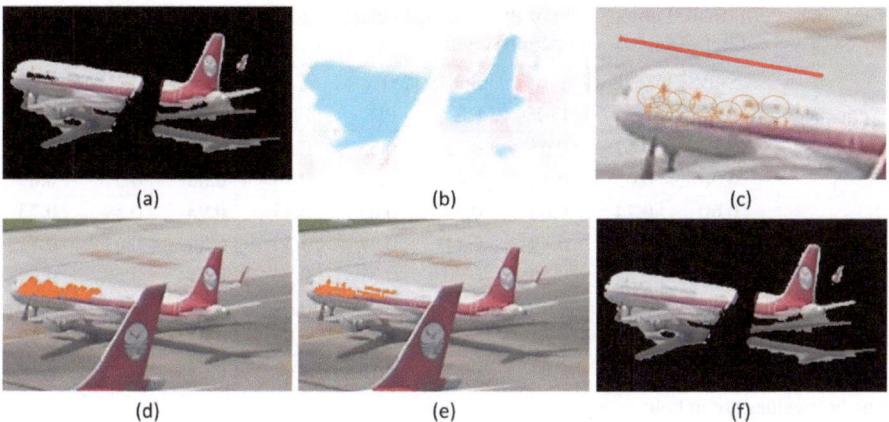

**Fig. 3.13** A specific example of sample synthesis and screening based on motion diffusion. Reprinted with permission from [21]

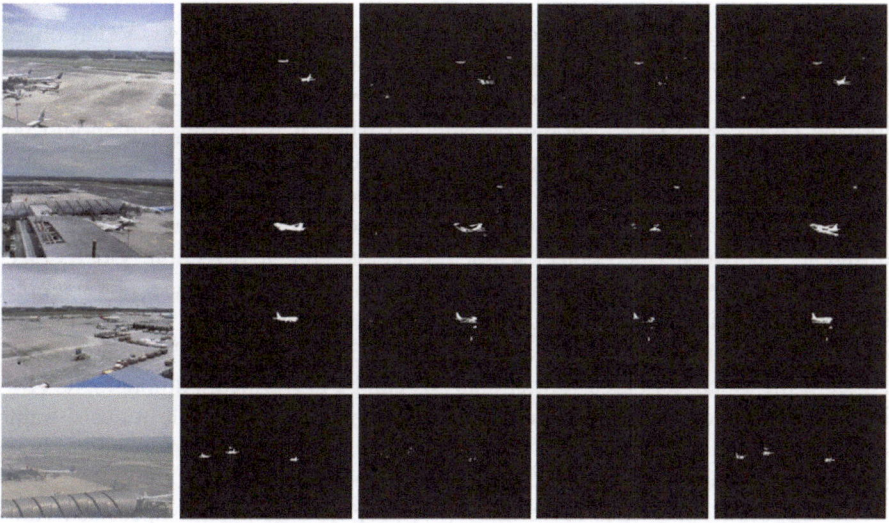

**Fig. 3.14** Comparison of three change detection algorithms on four sequences on the AGVS-S dataset. Adapted with permission from [21]

In Fig. 3.14, from left to right are the original images in AGVS-S, ground truth, segmentation results by KNN [22], PBAS [23], and our method, respectively. It can be seen that structural prior information does help improve the performance of change detection, which is further confirmed in Table 3.1. Eight sequences in AGVS-S are tested in Table 3.1. Three commonly used metrics, $RE$, $PR$, and $FM$, are used for performance comparison. These three indicators have been introduced in the previous chapter.

**Table 3.1** Performance comparison of three change detection algorithms on eight sequences on the AGVS-S dataset. Adapted with permission from [21]

| Sequences | SuBSENSE [24] | | | SOBS [25] | | | Proposed [21] | | |
|---|---|---|---|---|---|---|---|---|---|
| | RE | PR | FM | RE | PR | FM | RE | PR | FM |
| S16 | 0.52 | **0.87** | 0.65 | 0.22 | 0.55 | 0.31 | **0.87** | 0.68 | **0.76** |
| S22 | 0.58 | 0.65 | 0.62 | 0.08 | **0.67** | 0.14 | **0.86** | 0.57 | **0.69** |
| S3 | 0.60 | **0.74** | 0.63 | 0.33 | 0.63 | 0.43 | **0.93** | 0.59 | **0.73** |
| S12 | 0.43 | 0.57 | 0.49 | 0.41 | 0.28 | 0.33 | **0.84** | **0.61** | **0.71** |
| S1 | 0.55 | 0.54 | 0.55 | 0.20 | 0.31 | 0.25 | **0.81** | **0.58** | **0.68** |
| S14 | 0.35 | 0.46 | 0.40 | 0.03 | 0.01 | 0.02 | **0.86** | **0.54** | **0.66** |
| S15 | 0.64 | 0.59 | 0.61 | 0.43 | 0.26 | 0.33 | **0.90** | **0.62** | **0.73** |
| S18 | 0.01 | 0.31 | 0.01 | 0.01 | 0.13 | 0.01 | **0.65** | **0.72** | **0.68** |

The best values are in bold

### 3.2.3  Local-Global Feature Fusion Network

Video object segmentation is another binary segmentation problem. Video object segmentation also attempts to classify the pixels in each frame into two categories: foreground and background. In change detection, the foreground is defined as a moving target, so a target should be detected when it is moving and should not be detected when it is not moving. In video object segmentation, the foreground is manually specified in the first frame and should be detected in each subsequent frame, regardless of whether it is moving or not. So there are certain similarities and differences between video object segmentation and change detection. Video object segmentation is sometimes used in airport ground video surveillance. For example, when the administrator pays special attention to a certain aircraft target, the video object segmentation algorithm can be used to extract all the information of this target in the video, and then further analysis can be performed. Although AGVS-S is a change detection dataset, if only the targets with ground truth in every frame of the entire video are retained, then AGVS-S can also be used for video object segmentation research.

For airport ground video surveillance, we propose a new semi-supervised video object segmentation method: Local-Global Feature Fusion Network (LGFF-Net) [26]. This method can produce segmentation results in an end-to-end manner without online fine-tuning. LGFF-Net consists of three important modules: global encoder, local encoder, and joint decoder. The global encoder can extract scene-level features, which is conducive to ensuring the integrity of the segmentation results. The local encoder can extract the detail features of the target, which is conducive to improving the segmentation accuracy at the detail level. By using the global encoder and the local encoder together, a good balance can be found between completeness and accuracy. The method and experimental results are described as follows.

**Approach**  Figure 3.15 shows the flowchart of LGFF-Net. LGFF-Net contains three main modules, a global encoder, a local encoder, and a joint decoder. The global

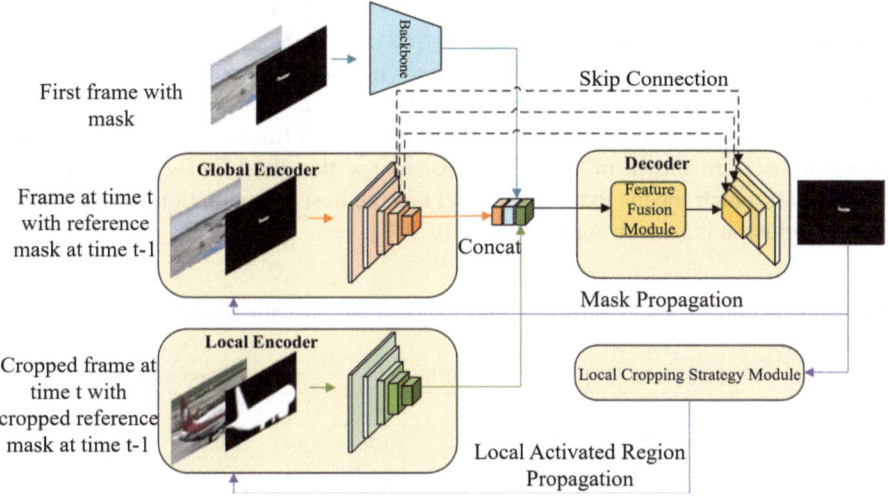

**Fig. 3.15** The flowchart of LGFF-Net. Adapted with permission from [26]

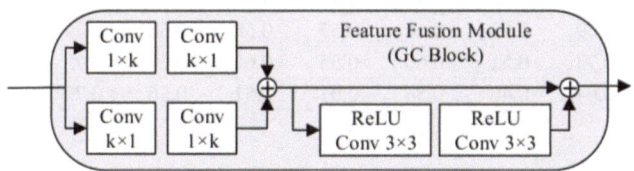

**Fig. 3.16** The structure of the feature fusion module. Reprinted with permission from [26]

encoder and the local encoder share the same parameters. The joint decoder consists of a feature fusion module and a decoder module. Skip connections are used in this module to synthesize shallow and deep features. The entire mask propagation process contains two loop steps.

The input of the global encoder and the local encoder is an image pair. The input image pair of the global encoder is the $t$ frame and the mask of the $t-1$ frame, while the input of the local encoder includes the output of the local cropping strategy module. The outputs of the two encoders will be connected with the first frame features extracted based on the backbone and fed into the joint decoder module. The feature fusion module in the joint decoder has the function of feature matching, and its network structure is shown in Fig. 3.16. We also designed a cropping strategy. Because for airport ground surveillance, the aircraft only exists in a small area of the entire airport ground. The cropping strategy helps to ignore the redundant background and focus on more details of the aircraft. This strategy can accelerate the learning process of the network and achieve finer segmentation. The decoder contains three refinement modules, and its input also includes skip connection features. The final mask output is obtained through the softmax layer.

**Experimental Results**  Five state-of-the-art video object segmentation algorithms are compared. They are Cascaded [29], FgSeg [30], FgSeg2 [31], RGMP [32], and SegFlow [33]. We use a new metric, the $IoU$ metric for performance comparison. The $IoU$ is the region similarity between the segmentation mask and the ground truth mask. The mean $IoU$ is the average of $IoU$ on all frames. The results in terms of mean $IoU$ are shown in Table 3.2. We can see that the LGFF-Net has the best mean $IoU$, which means that this method has the best segmentation completeness. This conclusion is further confirmed in Fig. 3.17.

**Table 3.2**  Performance comparison of video object segmentation algorithms on the AGVS-S dataset. Reprinted with permission from [26]

| Sequences | SuBSENSE [24] | | | SOBS [27] | | | Proposed [28] | | |
|---|---|---|---|---|---|---|---|---|---|
| | RE | PR | FM | RE | PR | FM | RE | PR | FM |
| S1 | 0.52 | 0.87 | 0.65 | 0.22 | 0.55 | 0.31 | 0.87 | 0.68 | 0.76 |
| S2 | 0.58 | 0.65 | 0.62 | 0.08 | 0.67 | 0.14 | 0.86 | 0.57 | 0.69 |
| S3 | 0.60 | 0.74 | 0.63 | 0.33 | 0.63 | 0.43 | 0.93 | 0.59 | 0.73 |
| S4 | 0.43 | 0.57 | 0.49 | 0.41 | 0.28 | 0.33 | 0.84 | 0.61 | 0.10 |
| S5 | 0.62 | 0.35 | 0.44 | 0.48 | 0.45 | 0.46 | 0.71 | 0.28 | 0.40 |
| S6 | 0.69 | 0.63 | 0.66 | 0.57 | 0.58 | 0.58 | 0.74 | 0.51 | 0.61 |
| S7 | 0.73 | 0.51 | 0.60 | 0.55 | 0.62 | 0.59 | 0.73 | 0.18 | 0.29 |
| S8 | 0.64 | 0.46 | 0.54 | 0.61 | 0.51 | 0.56 | 0.72 | 0.33 | 0.45 |

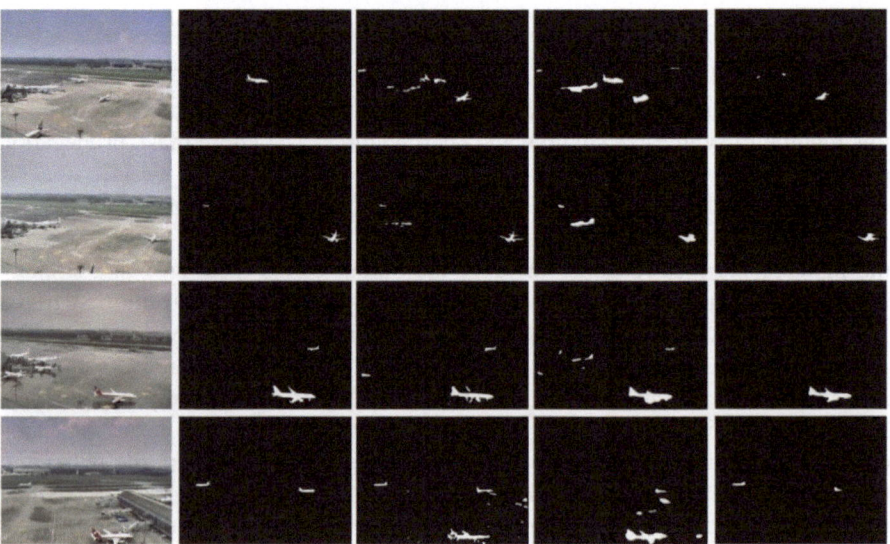

**Fig. 3.17**  Left to right: four frames from the AGVS-S dataset, ground truth, segmentation results by FgSeg [30], RGMP [32], and LGFF-Net, respectively. Adapted with permission from [26]

### 3.2.4 Color Prior Based Moving Object Detection

We already know that airplanes are difficult to segment completely. Because of the wingspan of airplanes, there are often serious segmentation defects in the wings and fuselage. In this section, we will show how to use another prior knowledge, color prior knowledge, to improve the accuracy of change detection. We noticed that part of the reason for the segmentation defects is the similarity of colors. Most airplanes are white, and the airport ground area where the airplanes are located also presents a similar gray-white tone, which can be seen from many of the previous airport image examples. When the airplane moves on the airport ground, it is easy to cause segmentation defects due to the similar color. The airport ground has such a color distribution because it is made of cement. Cement will have such a color after it dries. The ground of all airports is made of cement, so all airports have similar colors.

Since the color distribution of all airport grounds is similar, and this color interval is exactly where the aircraft segmentation defect is located, this inspires us to learn the color distribution of the airport ground and use it as prior knowledge to guide airport change detection. Based on this idea, we proposed a new change detection method [34], Scene-specific prior based Moving Object Detection (SMOD). In this method, we first use a bimodal Gaussian distribution to fit the color distribution of the airport ground. Then we design a cost function based on the bimodal Gaussian distribution and use this cost function within the framework of graph cut to impose classification bias, thereby reducing the probability of aircraft segmentation defect. This method and experimental results are described as follows.

**Approach** Figure 3.18 illustrates why we use a bimodal Gaussian distribution to fit the color distribution of the airport ground. Figure 3.18b is a grayscale histogram of Fig. 3.18a. It can be seen that Fig. 3.18b has an obvious bimodal distribution pattern, with two peaks corresponding to the white area and the gray area in the

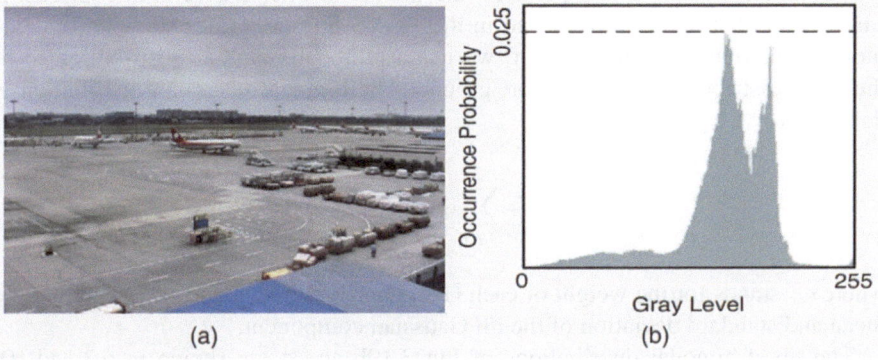

**Fig. 3.18** (**a**) One frame from the AGVS-S dataset. (**b**) Grayscale histogram of the airport ground area. Reprinted with permission from [34]

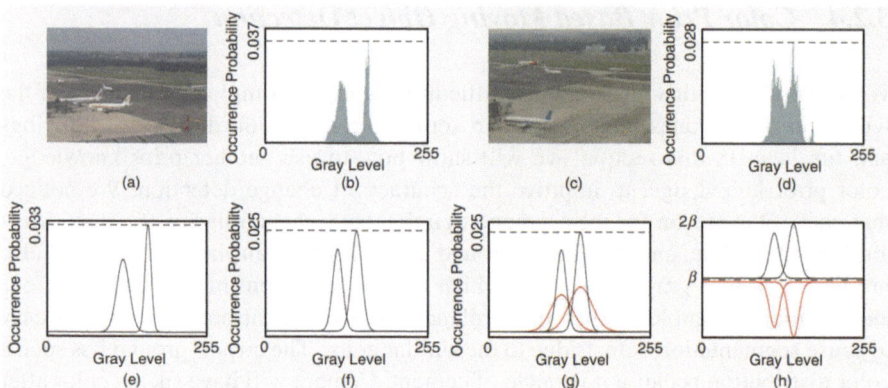

**Fig. 3.19** (**a**) and (**c**) Two frames from the AGVS-S dataset. (**b**) and (**d**) The gray histograms of (**a**) and (**c**). (**e**) and (**f**) The fitted dual-mode Gaussian distributions of (**b**) and (**d**), respectively. (**g**) The cost functions with different variances. (**h**) The two cost functions. Reprinted with permission from [34]

airport ground. We have also conducted a large number of experiments on the AGVS dataset series and found that most of the time this bimodal distribution pattern exists. Therefore, in the algorithm SMOD, we model the color pattern of the airport ground as a bimodal Gaussian distribution. However, in subsequent practical applications, we found that airports in different regions may not conform to the bimodal pattern due to different geographical factors and climatic conditions. At this time, a mixed Gaussian model can be used.

The fitting of color distribution and the construction of cost function are shown in Fig. 3.19. In Fig. 3.19, (a) and (c) are two frames from the AGVS-S dataset, and (b) and (d) are the grayscale histograms of (a) and (c), respectively. Please note that the grayscale histogram here is based on the statistics of the complete image, so it will be interfered by the non-airport ground area. However, because the airport ground area dominates in these two pictures, the impact of interference is not serious. It can be seen that the airport ground in the AGVS-S dataset does show an obvious bimodal distribution pattern. Next, we use a bimodal Gaussian distribution to fit the color distribution of the airport ground. The bimodal Gaussian distribution is defined as

$$p_h(z_i) = \sum_{r=1}^{2} \omega_r \mathcal{N}(z_i; \mu_r, \sigma_r), \tag{3.1}$$

where $\omega_r$ stands for the weight of each Gaussian component, and $\mu_r$ and $\sigma_r$ are the mean and standard deviation of the $i$th Gaussian component.

The fitted bimodal distributions of Fig. 3.19b and d are shown in (e) and (f). We already know that part of the reason for the aircraft detection defects is the similarity in color between the aircraft and the ground. Therefore, this bimodal distribution actually also reflects the detection defect probability of pixels with different grayscales. Now that the probability of detection defects is known, it can be used as prior knowledge to compensate for the detection defects. For example, pixels with the largest probability of detection defects receive the largest compensation, and pixels with smaller probability of detection defects receive small compensation. The core issue of this idea is, for different detection defect probabilities, what is the compensation level? If the compensation is too small, it may not work, and if the compensation is too much, it may also bring side effects. Therefore, we need a means to adjust the compensation level. Based on this consideration, we define another prior model,

$$p_h(z_i|\sigma) = \sum_{r=1}^{2} \omega_r \mathcal{N}(z_i; \mu_r|\sigma), \qquad (3.2)$$

where $\omega_r$ and $\mu_r$ are the same as in Eq. (3.1), and $\sigma$ is a variable parameter, which is used to control the compensation level. As shown in Fig. 3.19g, when $\sigma$ changes, the value of the prior model will also change, so the effect of controlling the compensation level can indeed be achieved. $\sigma$ is an important parameter in the SMOD algorithm and needs to be set carefully.

The prior model does not match the graph cut-based change detection model, so we designed a cost function based on the prior model, as shown in Fig. 3.19h. It can be seen that there are actually two cost functions, corresponding to the foreground likelihood term and the background likelihood term in the graph cut model. Based on the cost function, classification bias can be imposed in the graph model to reduce segmentation defects. For details of the cost function and graph model, please refer to the paper [34].

Please note that in Figs. 3.18 and 3.19, although the grayscale histogram of the original image shows an obvious bimodal distribution, it is actually still interfered by areas outside the airport ground area. It is only because the airport ground occupies a dominant position in these two examples that the interference is not serious. However, when the airport ground area does not occupy a dominant position, the grayscale histogram of the entire image is not necessarily bimodal, as shown in Fig. 3.20b. Considering this situation, our algorithm has an initialization step before running. The initialization task is to manually outline the airport ground area to eliminate interference from other areas. Figure 3.20a shows the result of manual outlining, and the corresponding grayscale histogram is shown in Fig. 3.20c. It can be seen that manual initialization can effectively remove the influence of interference.

**Experimental Results** The comparison between our method and SOBS [25], PBAS [23], FGMM [35], and ViBe [27] is shown in Figs. 3.21 and 3.22, respectively. It can be seen that color prior information can bring obvious performance improvement.

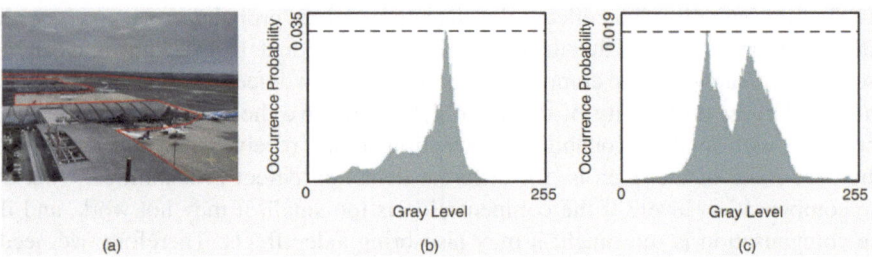

**Fig. 3.20** (**a**) A frame from AGVS-S. (**b**) and (**c**) The gray histograms of the whole image and the marked area, respectively. Reprinted with permission from [34]

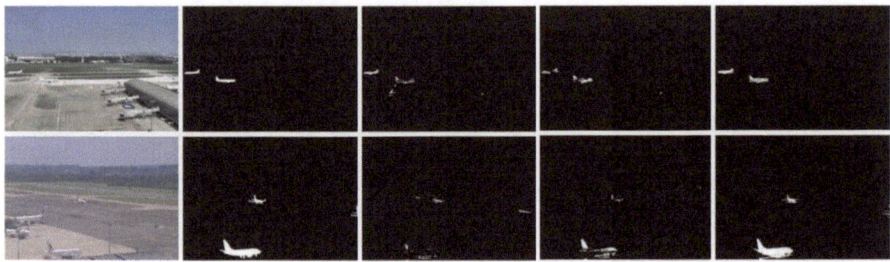

**Fig. 3.21** Left to right: two frames from AGVS-S dataset, ground truth, results by SOBS [25], PBAS [23], and SMOD, respectively. Reprinted with permission from [34]

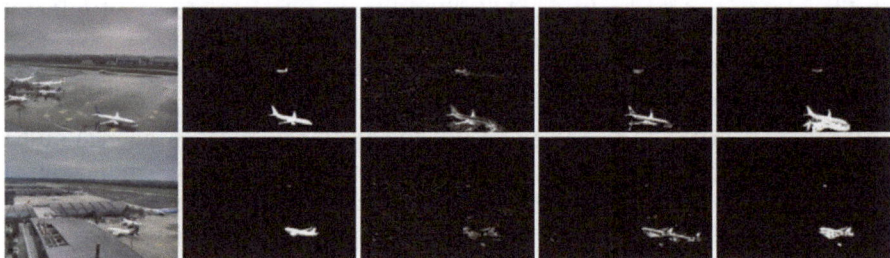

**Fig. 3.22** Left to right: two frames from AGVS-S dataset, ground truth, results by FGMM [35], ViBe [27], and SMOD, respectively. Reprinted with permission from [34]

### 3.2.5  ADS-B-Based Spatio-Temporal Alignment Network

We already know that the performance of video object segmentation in airport scenes is poor, and we have proposed a new method for airport video object segmentation that combines global and local features, which improves the segmentation performance to a certain extent. In this section, we try to use a satellite positioning information to further improve the performance of airport video object segmentation. Existing algorithms regard video object segmentation as a mask reconstruction process, propagating the mask frame by frame from the first frame to the last frame. However, due to object deformation and displacement, the errors in the mask propagation process will accumulate frame by frame, eventually leading to segmentation errors. Since the aircraft in the airport scene may have large appearance changes (such as when turning) and scale changes (such as when going from near to far), the phenomenon of propagation error accumulation is more obvious.

We propose an ADS-B-based Spatio-Temporal Alignment NAetwork (STA-Net) that uses ADS-B information to improve the performance of airport video object segmentation [36]. We have introduced ADS-B information in the previous section, which is a satellite positioning signal that can provide real-time position information of aircraft. In STA-Net, there is a spatio-temporal alignment step, which normalizes all targets in the historical mask to their current position based on ADS-B information, thereby generating multiple candidate masks. These candidate masks have compensated for the displacement of the target and can reduce the error accumulation to a certain extent. Furthermore, we select the mask that is most similar to the current moment as the final reference mask from all candidate masks, thereby compensating for the deformation. Since ADS-B information is real time, the reference mask can also be continuously updated, thereby preventing long-term error accumulation. The approach and experimental results of STA-Net [36] are described as follows.

**Approach** Let us first give a brief introduction to the ADS-B signal. The principle of ADS-B is shown in Fig. 3.23. First, the aircraft obtains its real-time coordinates based on the Global Navigation Satellite System (GNSS). Then, the transponder installed on the nose of the aircraft broadcasts the aircraft's physical coordinate information, speed, call sign, and other information in the form of ADS-B messages, with a broadcast interval of 0.5 seconds. After the airport ground station receives the ADS-B message, it will be transmitted to the air traffic control center, and the air traffic controller can obtain the aircraft's location information. The position data in the ADS-B message is in WGS84 format, which can be converted into the image coordinate system and is roughly located at the touchdown point of the aircraft's

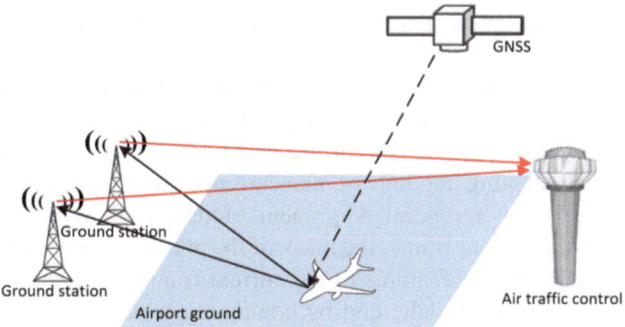

**Fig. 3.23** Illustration of the ADS-B signal. Reprinted with permission from [36]

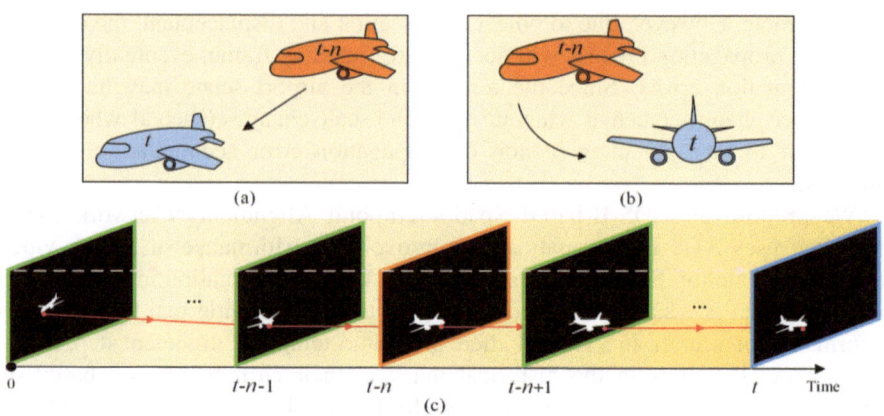

**Fig. 3.24** (**a**) Temporal misalignment, (**b**) spatial misalignment, and (**c**) principle of STA-Net. Top row: Orange and blue aircraft are at times $tn$ and $t$, respectively. Bottom row: Green and blue boxes represent candidate masks after normalization and current frame, respectively. Red nodes connected with arrows represent position information from ADS-B; orange box is the selected candidate mask for final inference. Reprinted with permission from [36]

front tires. The ADS-B information is used to guide the mask propagation process in STA-Net, and the principle is shown in Fig. 3.24.

The misalignment problem of airport video object segmentation is illustrated in Fig. 3.24a and b. Due to the displacement in Fig. 3.24a and appearance deformation of the aircraft in Fig. 3.24b, the reference frame and the current frame are spatio-temporally misaligned and finally result in error accumulation. In Fig. 3.24c, based on the ADS-B trajectory, all historical masks are normalized to the current frame to compensate the temporal misalignment. Furthermore, the normalized mask that is mostly similar to the current frame is selected as the final reference mask, so as to compensate the spatial misalignment.

Figure 3.25 shows the flowchart of STA-Net, which includes two steps of processing the past frame and the current frame. There is an encoder for each past frame. The input of the encoder also includes the segmentation mask and the ADS-B position information. The first frame and its manual annotation are used as the reference frame and reference mask, respectively. Each encoder uses ResNet50 as a feature extractor, and the network weights are initialized from the pretrained ImageNet model. In order to avoid the huge amount of computation required for frame-by-frame decoding, we perform the above decoding operation once every $Z$ frames. The processing module of the current frame includes an encoder, a decoder, and the Spatio-Temporal Alignment Module (STAM). The input of the encoder includes the current frame, the mask of the previous frame, and the location information in ADS-B. The features of the current frame extracted by the encoder are input into the STAM module, and by comparing with the features of the past frame, the target deformation and displacement are compensated, thereby realizing the effective use of ADS-B information. Therefore, the STAM module is the core

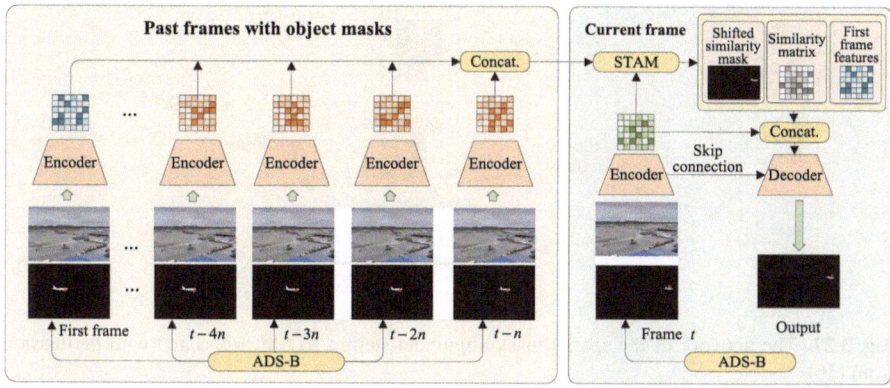

**Fig. 3.25** The framework of STA-Net, which includes two steps, processing the past frame and the current frame, respectively. Reprinted with permission from [36]

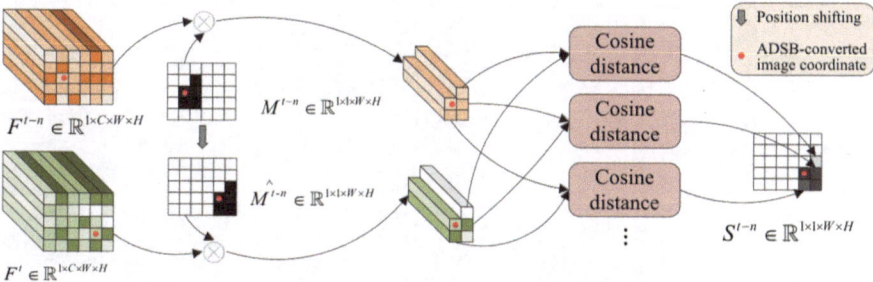

**Fig. 3.26** The network of the temporal mask alignment step in STAM. Reprinted with permission from [36]

module of this algorithm. By decoding the output of the STAM module, the final segmentation result is obtained. There are two main steps in STAM: temporal mask alignment and spatial mask alignment, which are shown in Figs. 3.26 and 3.27, respectively. For more details of the STAM module, please refer to [36].

**Experimental Results** Four state-of-the-art video object segmentation algorithms are selected for comparison. They are CascadedCNN [29], FgSeg [30], FgSeg2 [31], and SegFlow [33]. Experimental results are shown in Fig. 3.28. It can be seen that the STA-Net algorithm can obtain the most complete segmentation results. Furthermore, we designed an ablation experiment to compare the results of using and not using ADS-B information in STA-Net, as shown in Fig. 3.29. It can be seen that when ADS-B information is not used, the performance of STA-Net is comparable to that of RGMP [32], but after using ADS-B, it can bring obvious performance improvements.

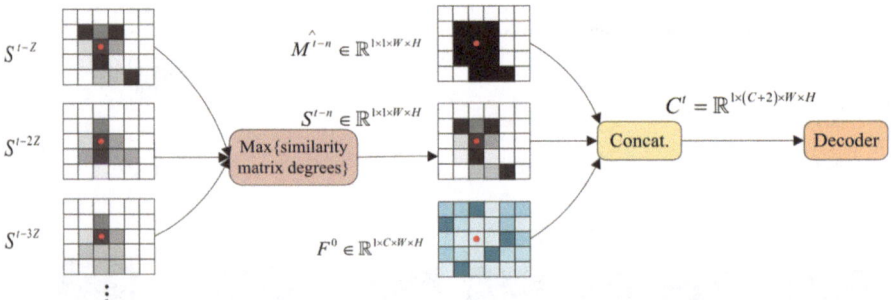

**Fig. 3.27** The network of the spatial mask alignment step in STAM. Reprinted with permission from [36]

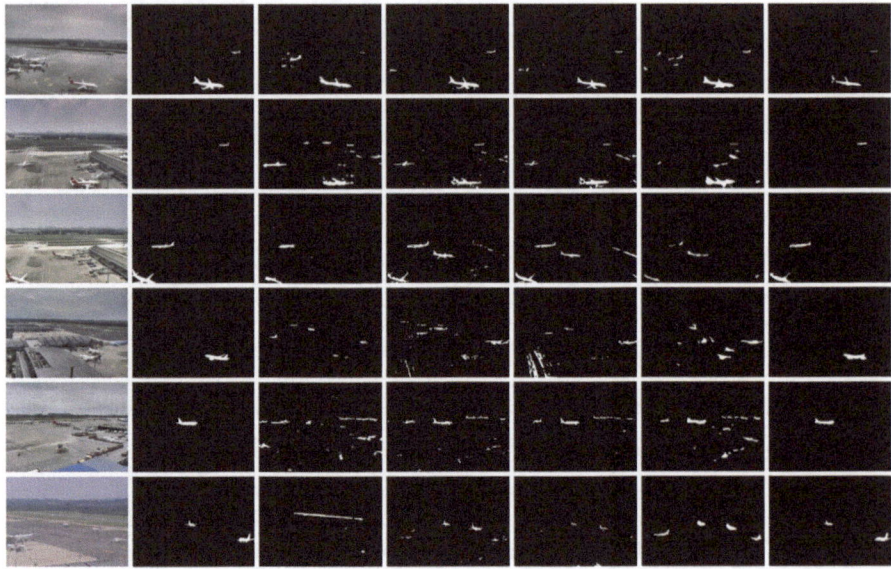

**Fig. 3.28** Top to bottom: typical frame from the AGVS-S dataset. Left to right: original frames, ground truth, segmentation results by CascadedCNN [29], FgSeg [30], FgSeg2 [31], SegFlow [33], and STA-Net, respectively. Reprinted with permission from [36]

## 3.3   Recognition

### 3.3.1   Related Work

Object recognition, also known as object detection, is a key technology in computer vision that involves accurately identifying and locating various objects in images or videos while assigning category labels. Since the success of AlexNet in 2012, object recognition has evolved from traditional techniques to deep learning methods.

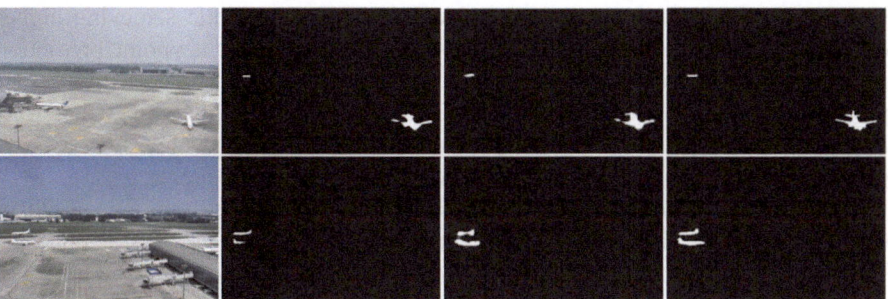

**Fig. 3.29** Left to right: two frames from AGVS-S, results by RGMP [32], and by STA-Net without and with ADS-B, respectively. Reprinted with permission from [36]

Today, it plays a crucial role in various applications, including video surveillance, autonomous driving, and medical image analysis.

In 2014, the proposal of R-CNN [37] marked the rise of two-stage object detection. R-CNN first uses Selective Search to generate a large number of candidate regions, then uses Convolutional Neural Networks (CNN) to extract features from each candidate region, and finally uses Support Vector Machines (SVM) for object classification. SPPNet [38] avoids repeated calculation of convolution features by introducing a spatial pyramid pooling layer that generates a fixed-length representation of a candidate region frame of arbitrary size. Fast R-CNN [39] takes the whole image as input, extracts the feature map by CNN and uses ROI pooling layer on the feature map to obtain the feature representation of each candidate region, and then uses fully connected layer for object classification and location regression. Faster R-CNN [40] introduces the Region Proposal Network (RPN) to unify the generation of region proposals with object classification and location into one model. In order to solve the problem that it is difficult to accurately detect both small and large targets with a single-scale feature map, FPN [41] realizes multi-scale feature representation of the input image by constructing a feature pyramid structure, which significantly improves the detection performance of the algorithm for targets of various scales. FPN does not significantly increase the computational complexity, and thus it is widely used in various object detection algorithms. Compared to Faster R-CNN, Cascade R-CNN [42] further stacks several cascaded modules in the detector head and uses different IOU thresholds to train the cascade detectors, which improves the accuracy of the two-stage target detection algorithm to a new level.

Unlike two-stage object detection algorithms, YOLO [43–45] pioneered the paradigm of one-stage object detection algorithms by using an end-to-end approach to analyze the entire image while predicting object bounding boxes and category probabilities for each region. SSD [46] improves the detection accuracy of one-stage detectors by detecting objects at different scales on different layers of the network. RetinaNet [47] reshapes the standard cross-entropy loss by introducing a new loss function, focal loss, which allows the detector to focus more on hard, misclassified samples during training. Focal loss allows the one-stage detector to achieve comparable accuracy to the two-stage detector while achieving higher detection speed. CornerNet [48] treats object detection as a keypoint prediction problem, and after obtaining the keypoints, decouples and regroups the corner points using additional embedding information to form a bounding box. CenterNet [49] treats an object as a single point and regresses all its attributes, such as size, orientation, position, attitude, etc., based on a reference center point.

In recent years, Transformer [50] has profoundly impacted the entire field of deep learning, especially the field of computer vision. DETR [51] first applied the Transformer architecture to the field of object detection, discarding the common steps of anchor mechanism, non-maximal suppression, etc., and instead adopting the encoder-decoder structure of Transformer to directly predict the images of the object bounding boxes and their categories. Deformable DETR [42] saves computational resources by introducing a deformable attention module that focuses on only a portion of the sampling points instead of the global information, while the use of multi-scale feature maps enhances the model's attention to local details and improves the detection performance for small targets.

Conditional DETR [52] believes that the query of the original DETR consists of two parts: content query and spatial query. The spatial query is a unified feature that does not target specific detail information in the image, while the content query needs to learn both the spatial key and the content key, which makes it difficult for the model to converge. Therefore, Conditional DETR generates a unique spatial embedding for each query, and when fusing with the content, it no longer uses the summation form but the splicing form. Anchor DETR [53] believes that the query of the original DETR is not focused on a specific area of interest, which poses a challenge to optimization. Therefore, Anchor DETR defines the query as a two-dimensional anchor, and each anchor corresponds to multiple answers, so that more positive samples can be generated, allowing the model to converge faster.

### 3.3.2   Airport Moving Object Recognition Network

We already know that in airport ground surveillance, our primary focus is on moving targets. Then existing recognition algorithms focus on how to detect the target more accurately, without paying attention to the motion attributes of the target. This prompted us to develop a recognition algorithm that can distinguish between motion and stillness. Based on this consideration, our group proposed an Airport Moving Object Recognition network (AMORnet) [54]. AMORnet only recognizes moving targets, and compared with traditional recognition algorithms, the recognition accuracy is significantly improved. AMORnet borrows the idea of change detection. The basic principle of change detection is to generate a background image or background model that does not contain moving targets and then distinguish the foreground and background by comparing the current frame and the background model. In AMORnet, we design a subnetwork to generate similar background images and use this as a reference to identify moving targets.

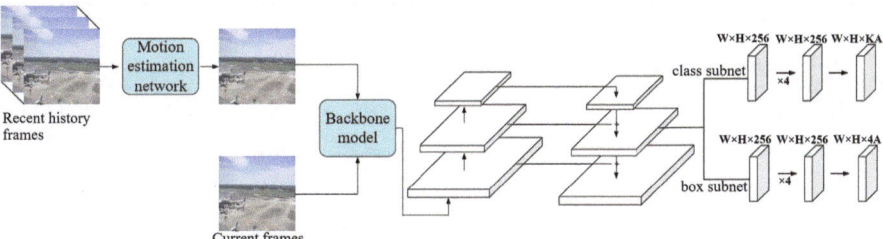

**Fig. 3.30**  Structure of the AMORnet. Reprinted with permission from [54]

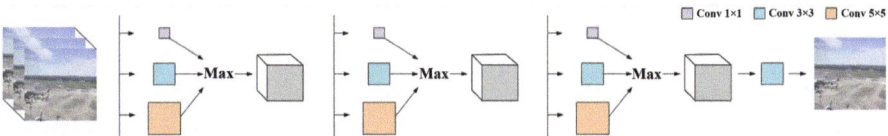

**Fig. 3.31**  Structure of the motion estimation network. Reprinted with permission from [54]

Because AMORnet is a coarse-grained recognition method, it is much faster than time-consuming segmentation algorithms such as change detection.

**Approach** AMORnet consists of a motion estimation network and a regression and classification module. The overall structure of AMORnet is shown in Fig. 3.30. Firstly, a clear and accurate background image is produced by the motion estimation network, as shown in Fig. 3.31. Comparisons of different background estimation methods are shown in Fig. 3.32. It can be seen that AMORnet can indeed obtain a relatively clean background estimation result. As shown in Fig. 3.30, the regression and classification module consists of a backbone network and two task-specific subnetworks. We use a multi-scale Feature Pyramid Network (FPN) as the backbone network, which can detect objects of different scales. At each layer of the feature pyramid feature map, the input feature map has 256 channels and 9 anchor sets, and the two task-specific subnetworks are used for regression and classification, respectively.

**Experimental Results** Because the AGVS-R dataset was not completed when the AMORnet algorithm was proposed, we used the AGVS-S dataset to test the AMORnet algorithm at that time. Although AGVS-S is a segmentation dataset, it can be easily converted into a moving object recognition dataset by calculating the bounding rectangle of each target in the ground truth. We use the first 10 videos in the AGVS-S dataset to train AMORnet, and the next 10 videos are used for testing. We use mAP as the evaluation metric. The mAP measures the average precision at different recall levels. The detection results of the proposed method with different numbers of samples to learn the background image are shown in Table 3.3. It can be concluded that the best results are achieved when the number of historical frames for the motion estimation network is 50. The comparison with the Faster RCNN method [40] is shown in Fig. 3.33. From this experiment, we can see that AMORnet can not

**Fig. 3.32** Comparison of different background estimation methods. (**a**) Four frames from the AGVS-S dataset, (**b**) and (**c**) results by temporal histogram-based method and the motion estimation network, respectively. Reprinted with permission from [54]

**Table 3.3** The mAP of AMORnet with different numbers of samples to learn the background image. Reprinted with permission from [54]

| | mAP | | | | |
|---|---|---|---|---|---|
| Number | 14 | 16 | 19 | 20 | Overall |
| 20 | 56.5 | 73.5 | 67.9 | 82.4 | 70.0 |
| 30 | 59.2 | 76.2 | **73.7** | 82.2 | 72.8 |
| 50 | **61.2** | 76.3 | 71.4 | **83.2** | **73.0** |
| 70 | 51.0 | **76.9** | 69.9 | 82.9 | 70.1 |

The best values are in bold

only distinguish between stationary and moving objects, but also detect occluded or incomplete objects.

### 3.3.3  ADS-B-Based Spatial-Temporal Object Detection Network

There are two main reasons for the low accuracy of aircraft recognition. First, an aircraft is a wingspan target that presents completely different appearances at different angles. In addition, the paint jobs of aircraft of different airlines are different. Second, unlike other surveillance scenarios, the airport area is very large, and the monitoring distance ranges from tens to hundreds or even thousands of meters, which often leads to the coexistence of extremely small scales and multiple scales. We already know that in airport scenarios, in addition to image data, there

**Fig. 3.33** Comparison of AMORnet with Faster RCNN [40]. (**a**) Frames from AGVS-S. (**b**) and (**c**) Results by Faster RCNN and AMORnet, respectively. Reprinted with permission from [54]

are also other modal data that can provide information that image data lacks. For example, ADS-B data can provide the four-dimensional position information (longitude, latitude, altitude, and time) of the aircraft, as well as some other additional information (weather, heading, aircraft model, etc.). Based on ADS-B data, we proposed the STA-Net algorithm, which can effectively improve the accuracy of video object segmentation.

In this section, we further consider using satellite positioning data such as ADS-B to improve the accuracy of aircraft recognition. For example, with the ADS-B coordinate as the center, there is a high probability that there are aircraft targets around, so ADS-B helps to improve the detection rate of target recognition. Because ADS-B can also provide the target's trajectory information, we can also consider using the temporal information provided by the ADS-B trajectory to improve the accuracy of video target recognition. Based on the above considerations, we propose an Airport Spatio-Temporal Network (ASTNet) for aircraft recognition [55], and its approach and experimental results are described as follows.

**Approach** ASTNet is a multi-scale spatio-temporal video object detection algorithm that leverages ADS-B data. It uses spatial and temporal information from ADS-B to enhance aircraft detection in videos. Spatially, it transforms single-

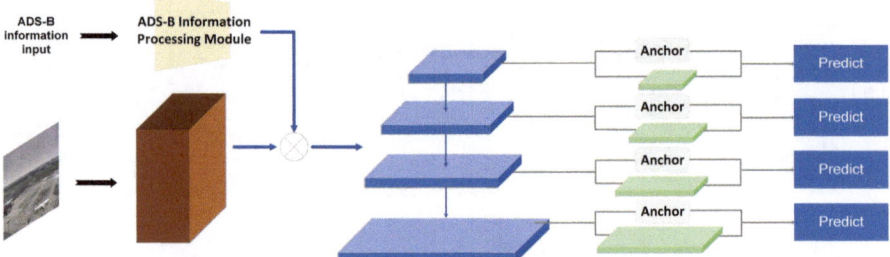

**Fig. 3.34** An overview of ASTNet. Reprinted with permission from [55]

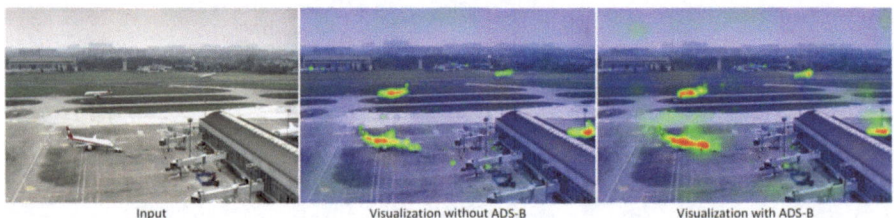

**Fig. 3.35** Left to right: a frame, aircraft features without feature enhancement, and aircraft features with ADS-B-based feature enhancement. Reprinted with permission from [55]

frame ADS-B signals into image point coordinates, constructs a Gaussian function centered on these coordinates, and applies spatial attention to amplify features of nearby targets. Temporally, it utilizes motion parameters from ADS-B trajectories, such as direction and scale, to match the optimal anchor aspect ratio, improving detection accuracy. The ASTNet flowchart, shown in Fig. 3.34, comprises two modules: the ADS-B data processing module and the ADS-B-based object detector. The processing module enhances object features using a Gaussian kernel based on ADS-B, while the detector predicts positions and aspect ratios from the ADS-B trajectory to generate the best matching anchor for final detection.

The feature enhancement in ADS-B data processing module is shown in Fig. 3.35. It can be seen that with the assistance of ADS-B signals, the features of the aircraft are indeed significantly enhanced. For details of the Gaussian kernel function used for feature enhancement, please refer to [55]. The principle of anchor generation in the object detector is illustrated in Fig. 3.36.

Figure 3.36 shows that there is a clear correspondence between the motion vector of the target and the ratio of its bounding box. Therefore, if we can obtain the motion vector of the target, we can predict the most suitable anchor for detecting the target, thereby improving the detection accuracy. The ADS-B trajectory provides exactly the motion information we want. The structure of the ADS-B-based anchor generation module is shown in Fig. 3.37. There are two branches in Fig. 3.37 to predict the location and ratio of the anchor, respectively. The complete network can be end-to-end trained.

Fig. 3.36 The principle of the anchor generation module. Reprinted with permission from [55]

Fig. 3.37 The anchor
generation module. Reprinted
with permission from [55]

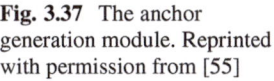

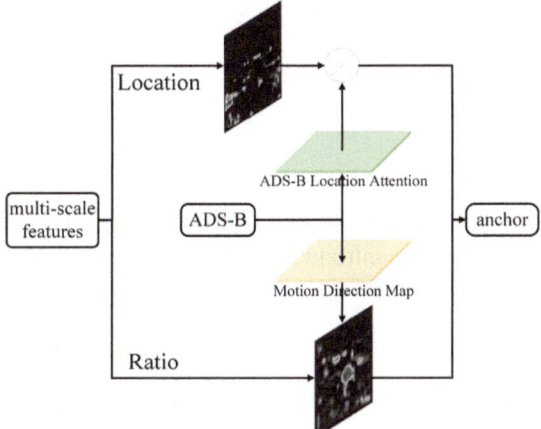

**Experimental Results** Because ASTNet is a video object detection algorithm, and the detection object is aircraft, a video object detection dataset containing aircraft is needed for experiments. There is currently no such dataset in the AGVS series we proposed. Our solution is to meet the requirements by modifying the AGVS-T dataset. AGVS-T is a video tracking dataset. Because tracking is based on recognition, the ground truth of AGVS-T contains recognition information. By extracting the recognition information of the aircraft target in AGVS-T, it can be converted into a video object detection dataset that meets the requirements. We train ASTNet for 12 epochs using NVIDIA 3060 TI (two images per GPU). For the training process, the 1x schedule represents 12 epochs. The initial learning rate is 0.001. The results of ASTNet on AGVS-T are shown in Fig. 3.38.

Furthermore, seven state-of-the-art algorithms [62] are selected for comparison. They are RetinaNet [47], PANet [56], FSSD [57], FCOS [58], YOLOX [59], GA-RPN [60], and DESTR [61]. The experimental results are shown in Table 3.4. It

**Fig. 3.38** The result of ASTNet on AGVS-T. Reprinted with permission from [55]

**Table 3.4** Comparison of object detection performance on AVGS-T dataset

| Network | Schedule | AP |
|---|---|---|
| RetinaNet [47] | 1x | 74.6 |
| PANet [56] | 1x | 75.2 |
| FSSD [57] | 1x | 79.6 |
| FCOS [58] | 1x | 77.3 |
| YOLOX [59] | 1x | 80.1 |
| GA-RPN [60] | 1x | 80.8 |
| DESTR [61] | 1x | 81.3 |
| Ours | 1x | **83.6** |

can be seen that the proposed method has the best Average Precision (AP), which shows that the introduction of ADS-B information does help improve the detection performance.

### 3.3.4   Full-Level Airport Scene Detection Network

Aircraft recognition is the core of recognition research for airport ground surveillance. Our experiments show that the recognition accuracy of medium-scale aircraft is higher, while the recognition accuracy of large-scale and small-scale aircraft is lower. Due to the scattered features of large-scale aircraft, the prediction box is difficult to accurately locate and can only cover part of the target. Small-scale aircraft are not rich in features themselves, and features are easily lost during feature extraction and fusion, resulting in frequent missed detection. We found that the low-level signals in the network contain more location and appearance information of small objects. Making full use of such low-level signals may help improve the detection performance of aircraft of all scales, including small-scale aircraft and large-scale aircraft. Therefore, we designed a full-level network for airport scenes to achieve this goal [63]. The approach and experimental results are described as follows.

Figure 3.39b shows the backbone of our method. CBAM [64] is an efficient attention module. CBAM obtains the result of feature refinement by mapping the

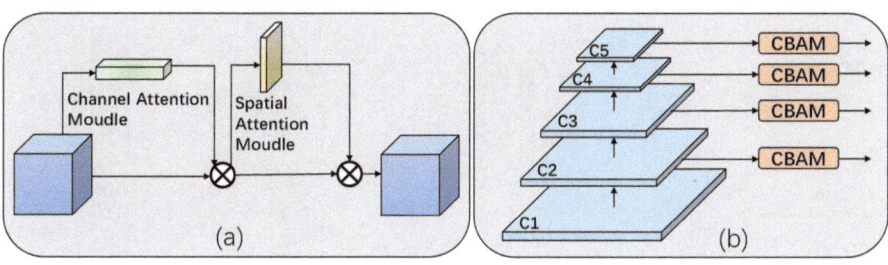

**Fig. 3.39** (**a**) shows the CBAM attention module; (**b**) shows the structure of backbone. Adapted with permission from [63]

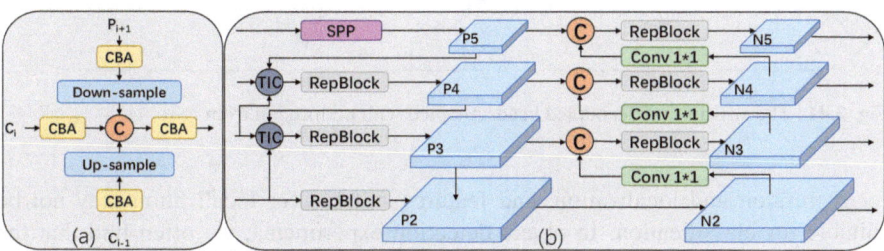

**Fig. 3.40** (**a**) shows the structure of Tri-directional Concatenation Module; (**b**) shows the structure of TiC-PAN. Adapted with permission from [63]

attention in the two dimensions of channel and space and multiplying it with the input feature map, as shown in Fig. 3.39a. The attention module allows the network to pay more attention to key features, thereby generally improving detection accuracy. We select ACON [65] with better nonlinear performance as the activation function in the convolution module. In addition, the backbone alternately uses two different sizes of convolution kernels and strides to downsample the features.

We developed a Tri-directional Concatenation module (TiC) to make use of low-level signals. TiC not only uses the local feature $C_i$ and the upper-level feature $P_{i+1}$, but also adds the lower-level feature $C_{i-1}$ during feature fusion, as shown in Fig. 3.40a. Inspired by FPN-PAN [66], TiC and FPN are combined to generate the TiC-FPN structure, as shown in Fig. 3.40b, where the neck part finally outputs four feature layers, corresponding to four different detection heads. The fourth detection head is located in the lower layer of the network. Although it increases the computational cost, it enhances the network's detection performance for small objects and improves positioning accuracy, which has been confirmed in subsequent experiments.

Object detection actually consists of two subtasks, classification and localization. Most detection algorithms use the same features to predict category and location information, that is, the network in the classification and regression heads shares parameters. However, the classification and localization tasks have different preferences for features. Features suitable for classification may not be suitable

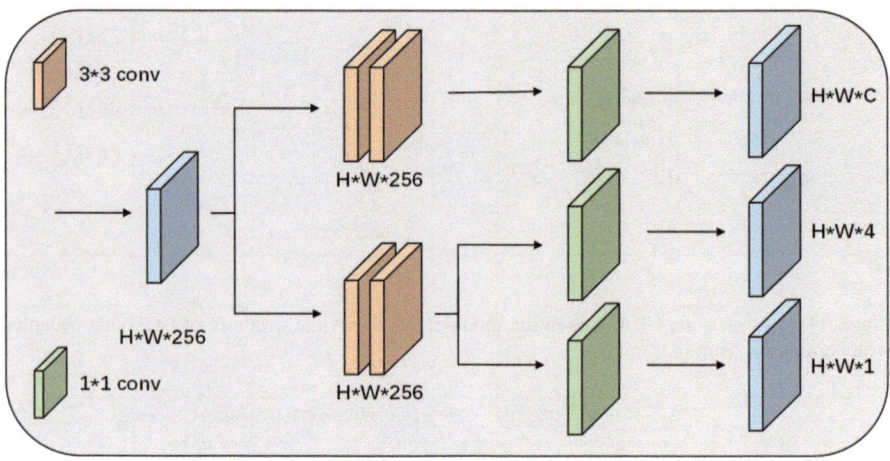

**Fig. 3.41** The structure of decoupled head. Adapted with permission from [63]

for regression and localization, and features suitable for localization may not be suitable for classification. In object detection experiments, we often find that the predicted box with the highest classification score is not necessarily the most accurate bounding box. Based on this consideration, we redesigned the head to decouple the features of the classification task and the localization task and added a separate IOU calculation branch to the localization branch, as shown in Fig. 3.41.

**Experiments Result** Because the AGVS-R dataset was not completed when this algorithm was proposed, we used the AGVS-S dataset for experiments. By calculating the bounding box of the aircraft target in the AGVS-S dataset, it can be easily converted into a recognition dataset. Totally nine state-of-the-art algorithms are selected for comparison. They are Faster RCNN [40], Cascade RCNN [67], CenterNet [49], YOLOv5s [68], RRNet [69], PP-YOLOE-s [70], DESTR [61], DPNetV3 [71], and YOLOX-S [59]. The experimental results are shown in Table 3.5. Ablation study is also carried out to verify the effectiveness of each module, as shown in Table 3.6. "w/o" is an abbreviation for "without," indicating that the network has removed that module.

It can be seen that the CBAM module can improve the detection accuracy without affecting the speed. The TiC module does improve the AP value, indicating that this algorithm does effectively utilize low-level information. The accuracy decreases after removing the fourth detection head, indicating that the C2 layer is useful for small object detection. Figure 3.42 shows more experimental results of our method on the AGVS-S dataset. It can be seen that the algorithm has good detection performance in different time periods such as daytime and dusk, as well as in scenes with densely parked aircraft.

**Table 3.5** Performance comparison of object detection on AGVS-S. Reprinted with permission from [63]

| Network | AP |
|---|---|
| Faster RCNN [40] | 63.2 |
| Cascade RCNN [67] | 68.9 |
| CenterNet [49] | 67.1 |
| YOLOv5s [68] | 78.6 |
| RRNet [69] | 80.1 |
| PP-YOLOE-s [70] | 79.2 |
| DESTR [61] | 79.8 |
| DPNetV3 [71] | 81.3 |
| YOLOX-S [59] | 80.2 |
| **Ours** | 82.1 |

**Table 3.6** Ablation study on AGVS-S. Reprinted with permission from [63]

| Network | AP | FPS |
|---|---|---|
| Baseline | 82.1 | 73 |
| Baseline w/o CBAM | 80.9 | 72 |
| Baseline w/o TiC | 80.3 | 70 |
| Baseline w/o extra head | 79.2 | 68 |

**Fig. 3.42** Some detection results on AGVS-S. Reprinted with permission from [63]

## 3.4 Tracking

### 3.4.1 Related Work

Tracking-By-Detection (TBD), also known as data association after detection, is the most commonly used strategy in Multiple Object Tracking (MOT). In recent years, coupled detection and tracking algorithms have also emerged, such as FairMOT [72], TrackFormer [73], etc. Regardless of the type of method, data association is the core of MOT. It first calculates the feature similarity between each track and all detection results and then determines the correspondence between tracks and detections based on different optimization strategies.

The most popular features used for similarity measurement are position, motion, and appearance. Position and motion similarity are accurate in short-distance matching, while appearance similarity is helpful in long-distance matching. For position, it mainly relies on the accuracy of the detector we mentioned in the previous section. For motion, a motion model can predict the potential position of the tracked object in the current frame, which is usually essential in MOT. For example, SORT [74] first adopts a Kalman filter (KF) to predict the position of the track in the new frame, and then calculates the intersection over union (IoU) between the detection box and the predicted box as the similarity. GIAOTracker [75] and StrongSORT [76] utilize variations of KF, especially Noise-Scale-Adaptive KF (NSA-KF), which integrates the detection score into KF to address low-quality detections and consider the detection noise scale. Tracktor [77] implements a simple camera motion compensation technique by aligning frames using image registration of moving camera sequences by maximizing the enhanced correlation coefficient. Finally, object discrimination and Re-identification (ReID) using deep appearance cues has become increasingly popular. These methods aim to enhance appearance features to improve inter-class variability for effective reassociation. For example, DeepSORT [78] uses a separate Re-ID model to extract appearance features as well as position and motion information from detection boxes to compute similarity metrics. Similarly, Track-RCNN [79] integrates a Re-ID head on top of Mask R-CNN [53] to regress bounding boxes and ReID features for each proposal.

After similarity measurement, an affinity matrix is obtained, which consists of all similarity values between trajectories and detections. Based on the optimization of the affinity matrix, the correspondence between trajectories and detections can be established. Therefore, the key element is the optimization strategy. Some methods use classic binary matching algorithms in graph theory, such as the Hungarian Algorithm (HA) and greedy allocation. These methods include SORT [74] and DeepSORT [78], as well as recent mainstream works [72, 76, 80, 81]. ByteTrack [82] uses a predefined bounding box confidence threshold to distinguish between low-confidence and high-confidence detection targets, thereby obtaining two separate affinity matrices and performing data association separately. The advantage of this is that low-confidence targets are not discarded, thus having better robustness to detection. Following this strategy, many SORT-based trackers [76, 81, 83, 83] have shown better performance. In addition, MOTDT [84] introduced a hierarchical data association strategy that utilizes IoU to associate objects when appearance features are unreliable.

There are some other methods for data association. MDP [85] uses Markov decision processes and reinforcement learning for online instance association. A hybrid data association framework [86] integrates the strengths of both online and offline methods for tracking multiple targets, employing a min-cost multi-commodity approach to correlate the same objects. RNN-LSTM [87] introduces a multi-object tracker that utilizes Recurrent Neural Networks (RNNs) for position prediction and Long Short-Term Memory (LSTM) for data association. Additionally, transformers represent another promising direction explored in works such as [73, 88–91]. In these methods, matching is conducted implicitly during the attention interaction process without relying on the HA. SGT [92] introduces a novel online graph tracker that utilizes higher-order relational features, enhancing discrimination by aggregating neighboring detections and their relationships.

### 3.4.2 Dual Markov Decision Processes for Airplane Tracking

We already know that there is a structural prior in airports. The layout of the airfields of all airports follows a similar pattern, including the clearance zone, the runway-taxiway zone, and the apron zone, as shown in Fig. 3.43. The clearance zone refers to the airspace above the airport ground, while the runway-taxiway zone is where aircraft accelerate for takeoff or landing. The apron zone connected to the runway-taxiway zone is used for taxiing and parking. Due to this structured layout, the motion patterns of aircraft on the airport ground are also predictable. Specifically, aircraft usually follow two directions of travel—straight and turning—and exhibit two speed patterns: constant speed and acceleration-deceleration. Based on this structure-motion prior information, we propose Dual Markov Decision Processes (DMDPs) for aircraft tracking [93]. Next, we first give the definition of the Markov decision process and then describe our method and experimental results.

A Markov decision process consists of the tuple $(S, \mathcal{A}, T(\bullet, \bullet), R(\bullet, \bullet))$:

- The target state $S_i \in S$ represents the state, e.g., motion state and tracking state, of each airplane.
- The action $A_i \in \mathcal{A}$ is a description of the behavior, e.g., from going straight to turning, of each airplane.
- The Markov state transition function $T(\bullet, \bullet) : S \times \mathcal{A} \to S$ describes the result of state transition by performing action $\mathcal{A}$ on state $S$.
- The reward function $R(\bullet, \bullet) : S \times \mathcal{A} \to R$ returns the cost after executing a certain action $\mathcal{A}$ on state $S$.

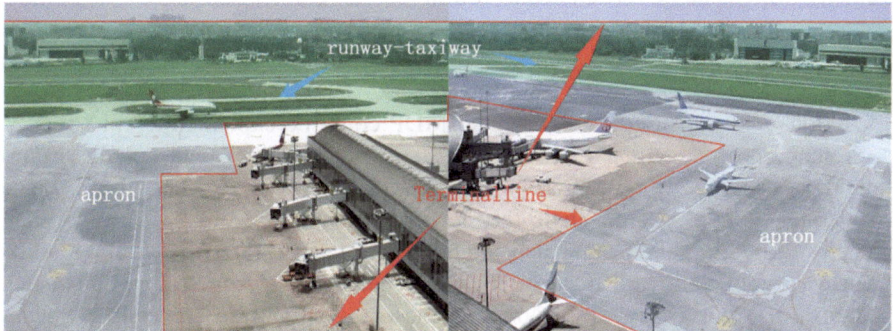

**Fig. 3.43** Illustration of the structure of the airport

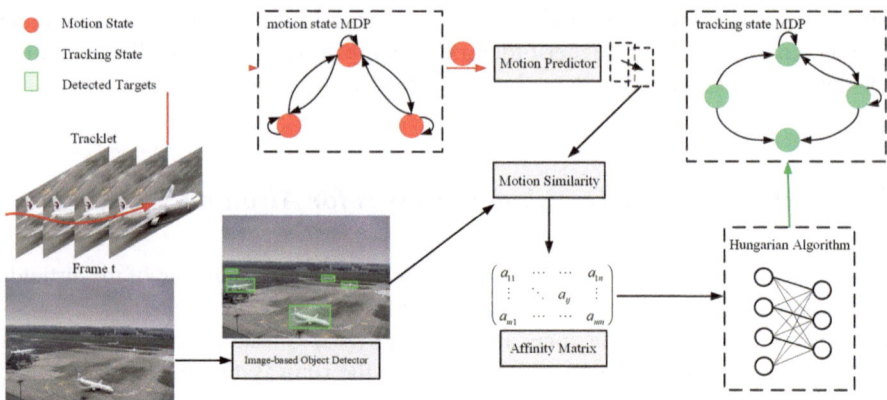

**Fig. 3.44** The framework of our two-level DMDPs. Reprinted with permission from [93]

**Approach** The proposed Dual Markov Decision Processes (DMDPs) method has a novel two-level MDP architecture, which can be formulated as follows:

$$S = S^M \cup S^T, \tag{3.3}$$

where $S^M$ is the motion state in motion state MDP (msMDP) and $S^T$ is the tracking state in tracking state MDP (tsMDP). A policy $\pi$ is a mapping from the state space $S$ to the action space $\mathcal{A}$, $\pi : S \to \mathcal{A}$. In other words, a policy determines what the next state is. Since the states of the first-level MDP are tracked and may be lost, the result of its policy is essentially the tracking result. The aim of decision-making in MDP is to identify a policy that maximizes total rewards. The framework of the two-level MDP is shown in Fig. 3.44.

The msMDP in detail consisted of the target state, action and transition function, and reward function. The motion state in msMDP is divided into three substates: Straight/Constant Velocity, Straight/Constant Acceleration, and Curve/Constant Velocity, which can be formulated as follows:

$$S^M = S^M_{S/CA} \cup S^M_{S/CV} \cup S^M_{C/CV}. \tag{3.4}$$

Actions and transition function of msMDP could be explained as this: Firstly, each state can maintain the existing state since the motion of the plane is continuous. Secondly, some states can switch to each other. When the airplane accelerates along a straight path, the state may transition from "straight/constant velocity" to "straight/constant acceleration." Similarly, when the airplane shifts from a straight path into a curve, the state can change from "straight/constant velocity" to "curve/constant velocity." These transitions are reversible, meaning "straight/constant acceleration" and "curve/constant velocity" can return to "straight/constant velocity." Furthermore, due to the airplane's significant inertia, it must maintain a constant speed before and after turning, which prevents any direct transition between "straight/constant acceleration" and "curve/constant velocity." There are seven potential transitions defined between the states of each target, corresponding to seven actions within the tracking status Markov decision processes. Based on the current state and the chosen action, the target can move to a new state, as outlined below:

$$\mu_t^{S_j^M} = T\left(\mu_{t-1}^{S_i^M}, a_{S_i^M, S_j^M}\right),$$

(3.5)

where $\mu_t^{S_j^M}$ represents the target's tracking state at time $t$ as $S_j^M$, while $\mu_{t-1}^{S_i^M}$ represents the target's tracking state at time $t-1$ as $S_i^M$. The term $a_{S_i^M}$, $S_j^M$ indicates the action applied to a target in the $S_i^M$ state, leading to a transition into the $S_j^M$ state. $T(\bullet, \bullet)$ denotes the Markov state transition function.

As for msMDP's reward function, we calculate the reward function to achieve the maximum reward value. The decisions are focused on the S/CA substate, C/CV substate, and S/CV substate. S/CA substate. In the S/CA substate, the decision is to transfer to the S/CV state or keep S/CA state. The state of each target is modeled as $x = [x, y, s, r, \dot{x}, \dot{y}, \dot{s}]^T$, $x$ and $y$ represent the horizontal and vertical pixel locations of the center of the target, while the scale $s$ and $r$ represent the scale (area) and the aspect ratio of the target bounding box, respectively. We calculate the reward function as follows:

$$R_{S/CA}\left(\mu_{t-1}^{S_i^M}, a_{S_i^M, S_j^M}\right) = \begin{cases} +\theta_{S/CA}, \dot{x} \geq \dot{x}_{thd} \| \dot{y} \geq \dot{y}_{thd} \\ -\theta_{S/CA}, \dot{x} < \dot{x}_{thd} \& \dot{y} < \dot{y}_{thd} \end{cases},$$

(3.6)

where $(\dot{x}_{thd}, \dot{y}_{thd})$ represent the specified thresholds, $\theta_{S/CA} = +1$ if $a_{S_i^M, S_j^M} = a_{S_{S/CA}^M, S_{S/CA}^M}$, and $\theta_{S/CA} = -1$ if $a_{S_i^M, S_j^M} = a_{S_{S/CA}^M, S_{S/CV}^M}$, as shown in Fig. 3.44.
C/CV substate. In the C/CV substate, the decision is whether to transition to the S/CV state or stay in the C/CV state. Due to the rigid structure of the airplane, its appearance changes dynamically during curved movements. To address this, we employ a deep neural network to extract the airplane's appearance features. We define the reward function for the C/CV state based on the representation of appearance features,

$$R_{C/CV}\left(\mu_{t-1}^{S_i^M}, a_{S_i^M, S_j^M}\right) = \begin{cases} +\theta_{C/CV}, E_{dis}(F_t, F_{t-\tau_{app}}) > d_{thd} \\ -\theta_{C/CV}, E_{dis}(F_t, F_{t-\tau_{app}}) \leq d_{thd} \end{cases},$$

(3.7)

where $d_{thd}$ is a specified threshold, $E_{dis}(\bullet, \bullet)$ represents the cosine distance, $F_t$ denotes the detected object's feature, and $F_{t-\tau_{app}}$ refers to the appearance feature from the $t - \tau_{app}$ frame on the trajectories. Here, $\tau_{app}$ is the time interval between the two frames used to calculate $E_{dis}(\bullet, \bullet)$. Additionally, $\theta_{C/CV} = +1$ if $a_{S_i^M, S_j^M} = a_{S_{C/CV}^M, S_{S/CV}^M}$, and $\theta_{C/CV} = -1$ if $a_{S_i^M, S_j^M} = a_{S_{C/CV}^M, S_{C/CV}^M}$, as illustrated in Fig. 3.44.

S/CV substate. In the S/CV substate, the msMDP must determine whether to keep the target in the S/CV state, transition it to an S/CA state, or switch it to a C/CV state. The solution for the reward function is comparable to the reverse process of $R_{S/CA}$ and $R_{C/CV}$.

As for motion modeling, the msMDP has three substates, each with a corresponding motion model. For S/CV and S/CA, state estimation is done using a KF and a variable parameter KF, respectively. Due to the airplane's rigid shape and dynamic appearance changes during turns, we use SiameseRPN [96] for motion prediction, an enhanced version of SiamesFC [97] with a Regional Proposal Network (RPN) for improved tracking of targets with changing appearances.

The tsMDP is mainly followed by reference [85]. The tracking state of an airplane is categorized into four substates: tentative, tracked, lost, and deleted, which can be defined as follows:

$$S^T = S^T_{tentative} \cup S^T_{tracked} \cup S^T_{lost} \cup S^T_{deleted}. \tag{3.8}$$

Actions and transition functions and reward functions of tsMDP could be shown like this: The transition model consists of four substates. The "tentative" state is the starting point for any trajectory. Once a target is detected by the pretrained object detector, it is assigned to the "tentative" state. If a target remains matched for several consecutive frames, it transitions to the "tracked" state; if not, it moves to the "deleted" state. A target in the "tracked" state can either remain tracked or switch to the "lost" state due to occlusion or moving out of view. A "lost" target can stay in the "lost" state, return to the "tracked" state if it reappears, or transition to the "deleted" state if it remains unseen for a specified period. The "deleted" state marks the end of the trajectory. The actions, transition functions, and reward functions we have developed follow the same principles [85].

**Experimental Results** Five state-of-the-art MOT algorithms are compared. They are SORT [74], CMOT [94], DeepSORT [78], MOTDT [84], and MDP [85]. Because we had not built the AGVS-T22 dataset when the DMDPs algorithm was proposed, we converted the change detection dataset AGVS-S into a tracking dataset and named it AGVS-T1, which was then used to test DMDPs and comparison algorithms. Five metrics are used for tracking performance comparison, MOTA, MOTP, IDF1, MT, and ML. They are defined as follows. MOTA (Multiple Object Tracking Accuracy) measures the overall tracking accuracy by considering false positives, false negatives, and ID switches.

$$\text{MOTA} = 1 - \frac{\text{FP} + \text{FN} + \text{IDSW}}{\text{GT}}. \tag{3.9}$$

MOTP (Multiple Object Tracking Precision) evaluates the precision by calculating the average distance between predicted and ground truth positions for matched objects.

$$\text{MOTP} = \frac{\sum_t \sum_i d_i}{C}. \tag{3.10}$$

IDF1 represents the F1 score for identity matching, balancing precision, and recall in terms of correct identity assignments.

$$\text{IDF1} = \frac{2 \times \text{IDTP}}{2 \times \text{IDTP} + \text{IDFP} + \text{IDFN}}. \tag{3.11}$$

MT (Mostly Tracked) is the ratio of ground truth objects tracked for more than 80% of their lifespan.

$$\text{MT} = \frac{\text{Mostly Tracked Objects}}{\text{Total Ground Truth Objects}}. \tag{3.12}$$

ML (Mostly Lost) is the ratio of ground truth objects tracked for less than 20% of their lifespan.

$$\text{ML} = \frac{\text{Mostly Lost Objects}}{\text{Total Ground Truth Objects}}. \tag{3.13}$$

We use two different object detectors, Faster R-CNN [40] and SDP [95], in combination with the DMDPs for evaluation. The experimental results are shown in Tables 3.7 and 3.8. It can be seen that DMDPs outperform other methods on the AGVS-T1 dataset with different detectors. More results by DMDPs are shown in Fig. 3.45. This figure shows that DMDPs are somewhat robust to the occlusion problem. Please note that in fact the AGVS-T1 dataset is not suitable for tracking. The videos in AGVS-T1 are all from the AGVS-S dataset, so it reflects the challenges faced by segmentation, but does not fully reflect the problems faced by tracking. For example, there are fewer cases of occlusion in the AGVS-T1 dataset. In subsequent research, we will use the AGVS-T22 dataset and its subsequent versions.

**Table 3.7** Tracking performance comparison on AGVS-T1 dataset. Best scores are marked in bold. Reprinted with permission from [93]

| Detector | Tracker | MOTA(%) ↑ | MOTP(%) ↑ | IDF1(%) ↑ | MT(%) ↑ | ML(%) ↓ |
|---|---|---|---|---|---|---|
| Faster R-CNN [40] | SORT [74] | 77.7 | 81.5 | 80.4 | 65.6 | 17.5 |
| Faster R-CNN [40] | CMOT [94] | 77.6 | 81.3 | 80.1 | 65.0 | 17.8 |
| Faster R-CNN [40] | DeepSORT [78] | 79.8 | 84.8 | 83.8 | 67.2 | 16.3 |
| Faster R-CNN [40] | MOTDT [84] | 81.7 | 88.2 | 85.6 | 69.0 | 14.9 |
| Faster R-CNN [40] | MDP [85] | 78.6 | 83.3 | 82.6 | 66.1 | 18.0 |
| Faster R-CNN [40] | Ours | **82.9** | **88.9** | **86.1** | **68.6** | **13.5** |

**Table 3.8** Tracking performance comparison on AGVS-T1 dataset. Best scores are marked in bold. Reprinted with permission from [93]

| Detector | Tracker | MOTA(%) ↑ | MOTP(%) ↑ | IDF1(%) ↑ | MT(%) ↑ | ML(%) ↓ |
|---|---|---|---|---|---|---|
| SDP [95] | CMOT [94] | 78.0 | 83.1 | 81.2 | 65.8 | 17.3 |
| SDP [95] | DeepSORT [78] | 80.3 | 85.3 | 84.6 | 68.0 | 15.6 |
| SDP [95] | MOTDT [84] | 82.5 | 88.5 | 86.3 | 69.4 | 14.3 |
| SDP [95] | MDP [85] | 79.3 | 83.5 | 83.3 | 66.8 | 17.4 |
| SDP [95] | Ours | **83.8** | **89.4** | **86.2** | **70.2** | **13.0** |

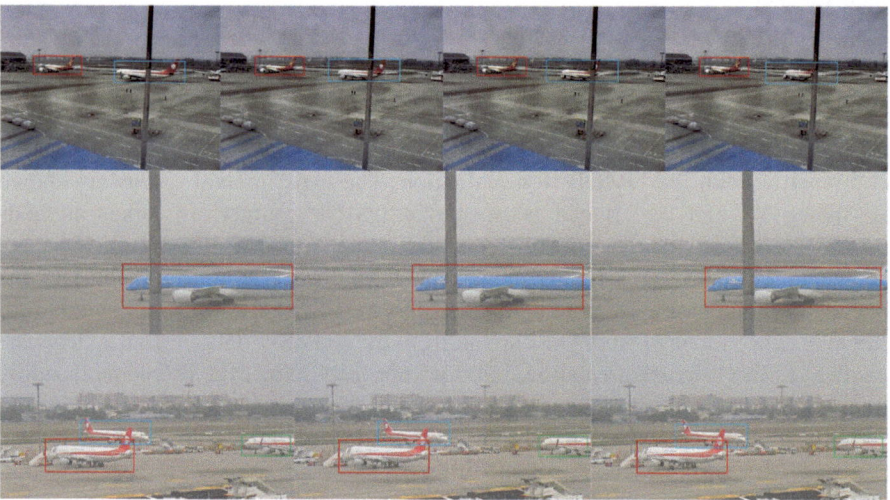

**Fig. 3.45** Some tracking results by DMDPs on AGVS-T1 dataset. The same identity is labeled by box with the same color. Adapted with permission from [93]

### 3.4.3   ADS-B-Based Multiple Object Tracking in Airport Ground

Although multiple object tracking (MOT) has achieved a great success, there is still a lack of effective solutions for multiple aircraft tracking. Besides common problems in tracking, there are some special issues in MAT, and among them, the multi-scale targets as well as similar color and shape between aircraft pose the greatest challenges. We have previously seen that satellite positioning information can effectively improve the performance of segmentation and recognition. In this section, we will propose a preliminary solution for multiple aircraft tracking based on ADS-B data.

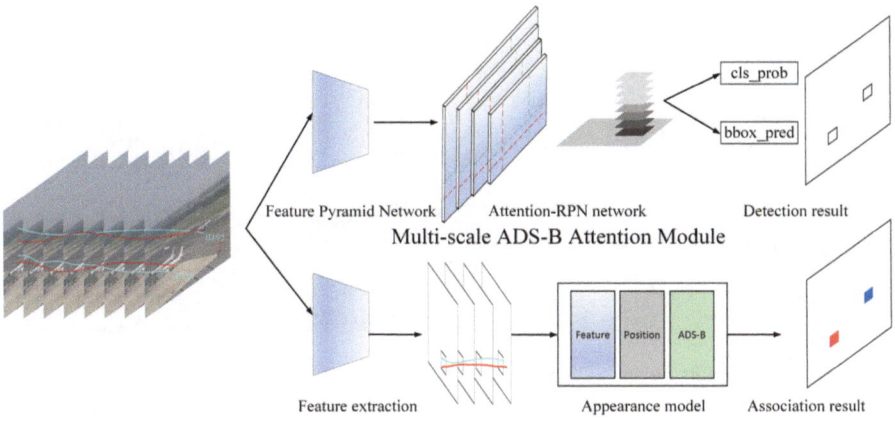

**Fig. 3.46** The flowchart of ADS-B-based multiple object tracking

**Approach** The flowchart of our preliminary approach is shown in Fig. 3.46. To make use of ADS-B, firstly an ADS-B-guided Object Detection module (AOD) is developed, which focuses on the regions indicated by ADS-B location information in multi-scale pyramid feature space. Miss-detection of small targets and false detection can be greatly reduced by AOD. Secondly, we build an ADS-B-guided Data Association module (ADA) that uses ADS-B trajectory as the reference of data association. ADS-B packages with the same ID naturally give the trajectory of an aircraft, that is, the ADS-B trajectory. Based on ADS-B trajectory, a novel appearance model is learned where the training samples are not necessarily from the video tracking trajectory but from the ADS-B trajectory. Furthermore, both the new and the traditional appearance models are competitively used to optimize ADA. Next, AOD and ADA are unified in an end-to-end network to complete the tracking task.

The main process of AOD is similar to the standard YOLOX [59], but there are two main differences. The first is the selection of anchor size, and the second is the classification enhancement of ADS-B-guided. Because there are aircraft of different sizes in airport ground video surveillance, the anchor size needs to be carefully designed. By analyzing all aircraft instances in the AGVS series, we redesigned the aspect ratio of the anchor to 2:3, 2:5, 2:7. In order to adapt to feature maps of different sizes, the anchor will be scaled according to the size of the feature map. The location of the ADS-B signal is on the front wheel of the aircraft and is just a point. We use the ADS-B coordinates as the center of attention and design a Gaussian kernel function to enhance the aircraft recognition capability within the kernel function coverage area.

The accuracy of the data association depends on the affinity matrix, including appearance affinity and spatial affinity. Therefore, we build an appearance model and a location prediction model based on the ADS-B trajectory to make the affinity

**Fig. 3.47** Some results by the ADS-B-based tracking solution

matrix more descriptive in ADA. Furthermore, considering that the aircraft detection in the airport is poor due to the small scale and other reasons, we adopted a two-level data association strategy similar to Bytetrack [82] to improve the tracking accuracy of small targets with low detection confidence.

**Experimental Results**   We combine AOD and ADA into an end-to-end model for joint learning. The backbone network of the model is DarkNet-53. Regarding the input samples of this model, it is divided into two categories. Images without ADS-B signals are learned in a standard way. Images containing the ADS-B signal, the threshold is adjusted on the feature map through position mapping to improve the model's attention to the ADS-B signal areas. To reduce over-fitting and increase the robustness of the model, some data enhancement techniques are applied. We train the model for 30 epochs using standard SGD and initialize the learning rate to $10^{-3}$ to reduce the loss. Some preliminary experimental results are shown in Fig. 3.47. It can be seen that satellite positioning information also helps to improve the tracking performance.

## 3.5   Other Algorithms

In addition to developing dedicated segmentation, recognition, and tracking algorithms, various preprocessing and postprocessing techniques can also be considered to improve image quality or refine algorithm results, thereby indirectly improving the performance of computer vision algorithms for airport ground surveillance. Here we take change detection as an example to discuss three common preprocessing techniques: white balance correction, image dehazing, and low-light image enhancement.

### 3.5.1  White Balance Correction

White balancing aims to normalize the lighting effects of the captured scene so that all objects appear as if they were captured under ideal "white light." White balancing is one of the first color processing steps applied to the raw RGB (rRGB) image from the sensor. After white balancing and some additional color rendering steps, the rRGB image is converted into a final standard RGB (sRGB) image. Most digital cameras provide automatic or adjustable white balance settings during image capture. Since the camera's own white balance settings may be incorrect, as well as for some aesthetic considerations, it is necessary to perform post-capture white balance correction. This task is challenging because the rRGB information is lost in the white balanced image. [100] proposed a sample-based framework to directly correct sRGB images captured with incorrect white balance settings, rather than recovering rRGB values. This work was further improved in deep-WB [98] by proposing a single deep learning framework to compare with the KNN strategy in [100]. Experimental results using deep WB as AGVS preprocessing are shown in Fig. 3.48. The improvement in detection performance in Fig. 3.48 is not significant, probably because white balance correction mainly affects the hue of the image rather than the contrast. However, given the importance of white balance, it is worthwhile to continue studying the relationship between white balance and airport video surveillance in the future.

### 3.5.2  Low-Light Image Enhancement

Nighttime videos is a huge challenge for many computer vision tasks. The difficulty of this problem lies in the low visibility or low contrast in videos due to the low-light imaging condition. We consider using low-light image enhancement as the preprocessing of change detection, and the experimental result is shown in Fig. 3.49. It can be seen that low-light image enhancement can improve the detection performance by increasing the image contrast. Some low-light enhancement methods focus on the under-exposure problem since it is the direct result of low light [102]. Both under-exposure and over-exposure are considered in [103] for image enhancement.

**Fig. 3.48** Left to right: one frame from AGVS-S, by white balance correction algorithm deep-WB [98], change detection by SegFlow [99] without and with deep-WB as preprocessing, respectively

**Fig. 3.49** Left to right: one frame from sequence V3 in AGVS-S dataset, by low-light image enhancement algorithm Zero-DCE [101], change detection by FgSegNet [30] without and with Zero-DCE as preprocessing, respectively

Another problem is video noise, which becomes significant due to low signal-to-noise ratio in nighttime videos. This problem is studied in [104] from the perspective of frequency domain. Unlike other learning-based methods, Zero-DEC [101] does not require any paired or unpaired data in the training process, thus avoiding the risk of over-fitting. Please note that nighttime videos have not been included in AGVS-S, because ground truth is not provided for the six sequences V1 to V6 in AGVS-S. This is because AGVS-S is a relatively early dataset that only considered daytime scenes. However, considering that the airport ground is a good scene for the research of image enhancement, for example, we can see over-exposure and under-exposure at the same time, we plan to add nighttime videos to the next version of AGVS-S. In later datasets, such as AGVS-T and AGVS-R, low-light scenes have been fully considered.

### 3.5.3   Image Dehazing

The haze dramatically degrades the visibility of outdoor videos and affects many high-level computer vision applications. The atmospheric scattering model describes the principle of image degradation due to haze:

$$I(x) = J(x)t(x) + A(1 - t(x)), \tag{3.14}$$

where $I(x)$ represents the observed hazy image, while $J(x)$ is the ideal, haze-free image. $A$ denotes the atmospheric illumination, and $t(x)$ is the transmission function that models the light attenuation as it travels from the scene to the camera. The goal of image dehazing is to recover $J(x)$ from the input image $I(x)$. To accomplish this, both the atmospheric light $A$ and the transmission matrix $t(x)$ must be estimated prior to performing dehazing. Instead of estimating $A$ and $t(x)$ separately, AOD-Net is the first one to jointly optimize $A$ and $t(x)$ within an end-to-end network [105]. Following this end-to end learning strategy, DCPDN is presented in [106], which has the property of edge preserving. The attention mechanism is introduced into FFA-Net [107], which performs well for regions with thick haze and rich texture details. The method in [108] formulates the image dehazing problem as the minimization

**Fig. 3.50** Left to right: one frame from AGVS-S, by dehazing algorithm AOD-Net [105], change detection by RGMP [32] without and with AOD-Net as preprocessing, respectively

of a variational model. The experimental result with AODNet as preprocessing of AGVS-S is shown in Fig. 3.50. It can be seen that the change detection performance is improved under haze weather, because image dehazing, like low-light image enhancement, can also increase the contrast of images.

# References

1. Long J, Shelhamer E, Darrell T (2015) Fully convolutional networks for semantic segmentation. In: 2015 IEEE/CVF Conference on Computer Vision and Pattern Recognition (CVPR). IEEE, Boston, pp 3431–3440
2. Zhao H, Shi J, Qi X, Wang X, Jia J (2017) Pyramid scene parsing network. In: 2017 IEEE/CVF Conference on Computer Vision and Pattern Recognition (CVPR). IEEE, Honolulu, Hawaii, pp 6230–6239
3. Zheng S, Lu J, Zhao H, Zhu X, Luo Z, Wang Y, Fu Y, Feng J, Xiang T, Torr PHS, Zhang L (2021) Rethinking semantic segmentation from a sequence-to-sequence perspective with transformers. In: 2021 IEEE/CVF Conference on Computer Vision and Pattern Recognition (CVPR). IEEE, Nashville, Tennessee, pp 6877–6886
4. Liu Z, Lin Y, Cao Y, Hu H, Wei Y, Zhang Z, Lin S, Guo B (2021) Swin transformer: hierarchical vision transformer using shifted windows. In: 2021 IEEE/CVF International Conference on Computer Vision (ICCV). IEEE, Montreal, pp 9992–10002
5. Chen L-C, Papandreou G, Kokkinos I, Murphy K, Yuille AL (2018) DeepLab: semantic image segmentation with deep convolutional nets, atrous convolution, and fully connected CRFs. IEEE Trans Pattern Anal Mach Intell 40(4):834–848. https://doi.org/10.1109/TPAMI.2017. 2699184
6. Lee J, Oh SJ, Yun S, Choe J, Kim E, Yoon S (2022) Weakly supervised semantic segmentation using out-of-distribution data. In: 2022 IEEE/CVF Conference on Computer Vision and Pattern Recognition (CVPR). IEEE, New Orleans, Louisiana, pp 16876–16885
7. Zhang D, Zhang H, Tang J, Hua X-S, Sun Q (2020) Causal intervention for weakly-supervised semantic segmentation. In: 2020 Advances in Neural Information Processing Systems (NIPS). MIT Press, Vancouver, pp 655–666
8. Liang Z, Wang T, Zhang X, Sun J, Shen J (2022) Tree energy loss: towards sparsely annotated semantic segmentation. In: 2022 IEEE/CVF Conference on Computer Vision and Pattern Recognition (CVPR). IEEE, Tel-Aviv, pp 16886–16895
9. Zhou B, Khosla A, Lapedriza A, Oliva A, Torralba A (2016) Learning deep features for discriminative localization. In: 2016 IEEE/CVF Conference on Computer Vision and Pattern Recognition (CVPR). IEEE, Las Vegas, pp 2921–2929
10. Dai J, He K, Sun J (2015) BoxSup: exploiting bounding boxes to supervise convolutional networks for semantic segmentation. In: 2015 IEEE International Conference on Computer Vision (ICCV). IEEE, Santiago, pp 1635–1643

11. Song C, Huang Y, Ouyang W, Wang L (2019) Box-driven class-wise region masking and filling rate guided loss for weakly supervised semantic segmentation. In: 2019 IEEE/CVF Conference on Computer Vision and Pattern Recognition (CVPR). IEEE, Long Beach, pp 3131–3140

12. Ouali Y, Hudelot C, Tami M (2020) Autoregressive unsupervised image segmentation. In: 2020 European Conference on Computer Vision (ECCV). Springer, Berlin, pp: 142–158

13. Ke R, Aviles-Rivero AI, Pandey S, Reddy S, Schönlieb CB (2022) A three-stage self-training framework for semi-supervised semantic segmentation. IEEE Trans Image Process 31:1805–1815. https://doi.org/10.1109/TIP.2022.3144036

14. He R, Yang J, Qi X (2021) Re-distributing biased pseudo labels for semi-supervised semantic segmentation: a baseline investigation. In: 2021 IEEE/CVF International Conference on Computer Vision (ICCV). IEEE, Montreal, pp 6910–6920

15. Ke Z, Qiu D, Li K, Yan Q, Lau RWH (2020) Guided collaborative training for pixel-wise semi-supervised learning. In: 2020 European Conference on Computer Vision (ECCV). Springer, Berlin, pp 429–445

16. Jain S, Wang X, Gonzalez JE (2019) Accel: a corrective fusion network for efficient semantic segmentation on video. In: 2019 IEEE/CVF Conference on Computer Vision and Pattern Recognition (CVPR). IEEE, Long Beach, pp 8858–8867

17. Sun G, Liu Y, Tang H, Chhatkuli A, Zhang L, Van Gool L (2022) Mining relations among cross-frame affinities for video semantic segmentation. In: 2022 European Conference on Computer Vision (ECCV). Springer, Tel-Aviv, pp 522–539

18. Lee SP, Chen SC, Peng WH (2021) GSVNET: guided spatially-varying convolution for fast semantic segmentation on video. In: 2021 IEEE International Conference on Multimedia and Expo (ICME). IEEE, Beijing, pp 1–6

19. Shelhamer E, Rakelly K, Hoffman J, Darrell T (2016) Clockwork convnets for video semantic segmentation. In: 2016 European Conference on Computer Vision (ECCV). Springer, Amsterdam, pp 852–868

20. Zhu X, Xiong Y, Dai J, Yuan L, Wei Y (2017) Deep feature flow for video recognition. In: 2017 IEEE Conference on Computer Vision and Pattern Recognition (CVPR). IEEE, Honolulu, Hawaii, pp 4141–4150

21. Zhang X, Wu H, Wu M, Wu C (2020) Extended motion diffusion-based change detection for airport ground surveillance. IEEE Trans Image Process 29:5677–5686. https://doi.org/10.1109/TIP.2020.2984854

22. Zivkovic Z, Van der Heijden F (2006) Efficient adaptive density estimation per image pixel for the task of background subtraction. Pattern Recogn Lett 27(7):773–780. https://doi.org/10.1016/j.patrec.2005.11.005

23. Hofmann M, Tiefenbacher P, Rigoll G (2012) Background segmentation with feedback: the pixel-based adaptive segmenter. In: 2012 IEEE/CVF Conference on Computer Vision and Pattern Recognition (CVPR). IEEE, Providence, Rhode Island, pp 38–43

24. St-Charles P-L, Bilodeau G-A, Bergevin R (2015) SuBSENSE: a universal change detection method with local adaptive sensitivity. IEEE Trans Image Process 24(1):359–373. https://doi.org/10.1109/TIP.2014.2378053

25. Maddalena L, Petrosino A (2012) The SOBS algorithm: what are the limits? In: 2012 IEEE/CVF Conference on Computer Vision and Pattern Recognition (CVPR). IEEE, Providence, Rhode Island, pp 21–26

26. Wu H, Li W, Wu M, Zhang X (2020) LGFF-Net: airport video object segmentation based on local-global feature fusion network. In: 2020 IEEE 2nd International Conference on Civil Aviation Safety and Information Technology (ICCASIT). IEEE, Weihai, pp 746–752

27. Barnich O, Van Droogenbroeck M (2009) ViBE: A powerful random technique to estimate the background in video sequences. In: 2009 IEEE International Conference on Acoustics, Speech and Signal Processing (ICASSP). IEEE, Taipei, Taiwan, pp 945–948

28. Kim K, Chalidabhongse TH, Harwood D, Davis L (2005) Real-time foreground–background segmentation using codebook model. Real-Time Imaging 11(3):172–185. https://doi.org/10.1016/j.rti.2004.12.004

29. Wang Y, Luo Z, Jodoin P-M (2017) Interactive deep learning method for segmenting moving objects. Pattern Recogn Lett 96:66–75. https://doi.org/10.1016/j.patrec.2016.09.014

30. Lim LA, Keles HY (2018) Foreground segmentation using a triplet convolutional neural network for multiscale feature encoding. Pattern Recogn Lett 112:256–262. https://doi.org/10.1016/j.patrec.2018.08.002

31. Lim LA, Keles HY (2020) Learning multi-scale features for foreground segmentation. Pattern Anal Applic 23(3):1369–1380. https://doi.org/10.1007/s10044-019-00845-9

32. Oh SW, Lee J-Y, Sunkavalli K, Kim SJ (2018) Fast video object segmentation by reference-guided mask propagation. In: 2018 IEEE/CVF Conference on Computer Vision and Pattern Recognition (CVPR). IEEE, Salt Lake City, Utah, pp 7376–7385

33. Cheng J, Tsai Y-H, Wang S, Yang M-H (2017) SegFlow: joint learning for video object segmentation and optical flow. In: 2017 IEEE/CVF International Conference on Computer Vision (ICCV). IEEE, Venice, pp 686–695

34. Zhang X, Qiao Y, Yang Y, Wang S (2023) SMod: scene-specific-prior–based moving object detection for airport apron surveillance systems. IEEE Intell Transp Syst Mag 15(1):58–69. https://doi.org/10.1109/MITS.2021.3122926

35. El Baf F, Bouwmans T, Vachon B (2008) Type-2 fuzzy mixture of gaussians model: application to background modeling. In: 2008 International Symposium on Visual Computing (ISVC). Springer, Berlin, Heidelberg, pp 772–781

36. Zhang X, Wang S, Wu H, Liu Z, Wu C (2022) ADS-B-Based spatiotemporal alignment network for airport video object segmentation. IEEE Trans Intell Transp Syst 23(10):17887–17898. https://doi.org/10.1109/TITS.2022.3160479

37. Girshick R, Donahue J, Darrell T, Malik J Rich (2014) Feature hierarchies for accurate object detection and semantic segmentation. In: 2014 IEEE/CVF Conference on Computer Vision and Pattern Recognition (CVPR). IEEE, Columbus, Ohio, pp 580–587

38. He K, Zhang X, Ren S, Sun J (2015) Spatial pyramid pooling in deep convolutional networks for visual recognition. IEEE Trans Pattern Anal Mach Intell 37(9):1904–1916. https://doi.org/10.1109/TPAMI.2015.2389824

39. Girshick R (2015) Fast R-CNN. In: 2015 IEEE/CVF International Conference on Computer Vision(ICCV). IEEE, Santiago, Chile, pp 1440–1448

40. Ren S, He K, Girshick R, Sun J (2017) Faster R-CNN: towards real-time object detection with region proposal networks. IEEE Trans Pattern Anal Mach Intell 39(6):1137–1149. https://doi.org/10.1109/TPAMI.2016.2577031

41. Lin T-Y, Dollar P, Girshick R, He K, Hariharan B, Belongie S (2017) Feature pyramid networks for object detection. In: 2017 IEEE/CVF Conference on Computer Vision and Pattern Recognition (CVPR). IEEE, Honolulu, Hawaii, pp 936–944

42. Zhu X, Su W, Lu L, Li B, Wang X, Dai J (2021) Deformable DETR: deformable transformers for end-to-end object detection. In: 2021 International Conference on Learning Representations(LCLR). Vienna, Austria

43. Redmon J, Divvala S, Girshick R, Farhadi A (2016) You only look once: unified, real-time object detection. In: 2016 IEEE/CVF Conference on Computer Vision and Pattern Recognition (CVPR). IEEE, Las Vegas, Nevada, pp 779–788

44. Redmon J, Farhadi A (2017) YOLO9000: better, faster, stronger. In: 2017/CVF IEEE Conference on Computer Vision and Pattern Recognition (CVPR). IEEE, Honolulu, Hawaii, pp 6517–6525

45. Redmon J, Farhadi A (2018) YOLOv3: an incremental improvement. In: 2018 IEEE/CVF Conference on Computer Vision and Pattern Recognition (CVPR). IEEE, Salt Lake City, Utah, pp 1–6

46. Liu W, Anguelov D, Erhan D, Szegedy C, Reed S, Fu C-Y, Berg AC (2016) SSD: single shot multibox detector. In: 2016 European Conference on Computer Vision (ECCV). Springer, Amsterdam, pp 21–37

47. Lin T-Y, Goyal P, Girshick R, He K, Dollar P (2017) Focal loss for dense object detection. In: 2017 IEEE/CVF International Conference on Computer Vision (ICCV). IEEE, Venice, pp 2980–2988

48. Law H, Deng J (2018) CornerNet: detecting objects as paired keypoints. In: 2018 European Conference on Computer Vision (ECCV). Springer, Munich, pp 734–750
49. Duan K, Bai S, Xie L, Qi H, Huang Q, Tian Q (2019) CenterNet: keypoint triplets for object detection. In: 2019 IEEE/CVF International Conference on Computer Vision (ICCV). IEEE, Seoul, Korea (South), pp 6568–6577
50. Vaswani A, Shazeer N, Parmar N, Uszkoreit J, Jones L, Gomez AN, Kaiser L, Polosukhin I (2017) Attention is all you need. In: 2017 Neural Information Processing Systems (NIPS). MIT Press, Long Beach, California, pp 21–25
51. Carion N, Massa F, Synnaeve G, Usunier N, Kirillov A, Zagoruyko S (2020) End-to-End object detection with transformers. In: 2020 European Conference on Computer Vision (ECCV). Springer, Glasgow, pp 213–229
52. Meng D, Chen X, Fan Z, Zeng G, Li H, Yuan Y, Sun L, Wang J (2021) Conditional DETR for fast training convergence. In: 2021 IEEE/CVF International Conference on Computer Vision (ICCV). IEEE, Montreal, Quebec, pp 3631–3640
53. Wang Y, Zhang X, Yang T, Sun J (2022) Anchor DETR: query design for transformer-based detector. Assoc Adv Artif Intell 36(3):2567–2575. https://doi.org/10.1609/aaai.v36i3.20158
54. Zhang Z, Zhang X, Chen D, Yu H (2022) Moving object recognition for airport ground surveillance network. In: 2022 International Conference on Mobile Networks and Management (MONAMI). Springer, Calafate, pp 335–343
55. Jiang L, Zhang X, Liu Y, Li T (2023) ADS-B-Based spatial-temporal multi-scale object detection network for airport scenes. In: 2023 International Conference on Image and Graphics (ICIG). Springer, Nanjing, pp 334–345
56. Wang K, Liew JH, Zou Y, Zhou D, Feng J (2019) PANet: few-shot image semantic segmentation with prototype alignment. In: 2019 IEEE/CVF International Conference on Computer Vision (ICCV). IEEE, Seoul, Korea (South), pp 9196–9205
57. Li Z, Yang L, Zhou F (2017) FSSD: feature fusion single shot multibox detector. arXiv:1712.00960, 2017
58. Tian Z, Shen C, Chen H, He T (2022) FCOS: a simple and strong anchor-free object detector. IEEE Trans Pattern Anal Mach Intell 44(4):1922–1933. https://doi.org/10.1109/TPAMI.2020.3032166
59. Ge Z, Liu S, Wang F, Li Z, Sun J (2021) YOLOX: exceeding YOLO series in 2021. arXiv:2107.08430, 2021
60. Wang J, Chen K, Yang S, Loy CC, Lin D (2019) Region proposal by guided anchoring. In: 2019 IEEE/CVF Conference on Computer Vision and Pattern Recognition (CVPR). IEEE, Long Beach, California, pp 2960–2969
61. He L, Todorovic S (2022) DESTR: object detection with split transformer. In: 2022 IEEE/CVF Conference on Computer Vision and Pattern Recognition (CVPR). IEEE, New Orleans, Los Angeles, pp 9367–9376
62. Chen K, Wang J, Pang J, Cao Y, Xiong Y, Li X, Sun S, Feng W, Liu Z, Xu J, Zhang Z, Cheng D, Zhu C, Cheng T, Zhao Q, Li B, Lu X, Zhu R, Wu Y, Dai J, Wang J, Shi J, Ouyang W, Loy CC, Lin D (2019) MMDetection: Open MMLab detection toolbox and benchmark. arXiv:1906.07155, 2019
63. Li X, Zhang X, Meng J, Jiang L (2023) A full-level based network to detect every aircraft in airport scene. In: International Conference on Image and Graphics (ICIG). Springer, NanJing, pp 212–223
64. Woo S, Park J, Lee J-Y, Kweon IS (2018) CBAM: convolutional block attention module. In: 2018 European Conference on Computer Vision (ECCV). Springer, Munich, pp 3–19
65. Ma N, Zhang X, Liu M, Sun J (2021) Activate or not: learning customized activation. In: 2021 IEEE/CVF Conference on Computer Vision and Pattern Recognition (CVPR). IEEE, Nashville, Tennessee, pp 8028–8038
66. Liu S, Qi L, Qin H, Shi J, Jia J (2018) Path aggregation network for instance segmentation. In: 2018 IEEE/CVF Conference on Computer Vision and Pattern Recognition (CVPR). IEEE, Salt Lake City, Utah, pp 8759–8768

67. Cai Z, Vasconcelos N (2018) Cascade R-CNN: delving into high quality object detection. In: 2018 IEEE/CVF Conference on Computer Vision and Pattern Recognition (CVPR). IEEE, Salt Lake City, Utah, pp 6154–6162

68. Jocher G, Chaurasia A, Qiu J (2023). Ultralytics YOLO (Version 5.0.0) [Computer software]. https://github.com/ultralytics/ultralytics

69. Chen C, Zhang Y, Lv Q, Wei S, Wang X, Sun X, Dong J (2019) RRNet: A hybrid detector for object detection in drone-captured images. In: 2019 IEEE/CVF International Conference on Computer Vision Workshop (ICCVW). IEEE, Seoul, Korea (South), pp 100–108

70. Xu S, Wang X, Lv W, Chang Q, Cui C, Deng K, Wang G, Dang Q, Wei S, Du Y, Lai B (2022) PP-YOLOE: an evolved version of YOLO. arXiv:2203.16250, 2022

71. Shi H, Zhou Q, Ni Y, Wu X, Latecki LJ (2022) DPNet: dual-path network for real-time object detection with lightweight attention. In: 2022 IEEE International Conference on Image Processing (ICIP). IEEE, Bordeaux, pp 771–775

72. Zhang Y, Wang C, Wang X, Zeng W, Liu W (2021) FairMOT: on the fairness of detection and re-identification in multiple object tracking. Int J Comput Vis 129(11):3069–3087. https://doi.org/10.1007/s11263-021-01513-4

73. Meinhardt T, Kirillov A, Leal-Taixe L, Feichtenhofer C (2022) TrackFormer: multi-object tracking with transformers. In: 2022 IEEE/CVF Conference on Computer Vision and Pattern Recognition (CVPR). IEEE, New Orleans, Louisiana, pp 8834–8844

74. Bewley A, Ge Z, Ott L, Ramos F, Upcroft B (2016) Simple online and realtime tracking. In: 2016 IEEE International Conference on Image Processing (ICIP). IEEE, Phoenix, Arizona, pp 3464–3468

75. Du Y, Wan J, Zhao Y, Zhang B, Tong Z, Dong J (2021) GIAOTracker: a comprehensive framework for MCMOT with global information and optimizing strategies in VisDrone 2021. In: 2021 IEEE/CVF International Conference on Computer Vision Workshops (ICCVW). IEEE, Montreal, Quebec, pp 2809–2819

76. Du Y, Zhao Z, Song Y, Zhao Y, Su F, Gong T, Meng H (2023) StrongSORT: make DeepSORT great again. IEEE Trans Multimedia 25:8725–8737. https://doi.org/10.1109/TMM.2023.3240881

77. Bergmann P, Meinhardt T, Leal-Taixe L (2019) Tracking without bells and whistles. In: 2019 IEEE/CVF International Conference on Computer Vision (ICCV). IEEE, Seoul, pp 941–951

78. Wojke N, Bewley A, Paulus D (2017) Simple online and realtime tracking with a deep association metric. In: 2017 IEEE International Conference on Image Processing (ICIP). IEEE, Beijing, pp 3645–3649

79. Voigtlaender P, Krause M, Osep A, Luiten J, Sekar BBG, Geiger A, Leibe B (2019) MOTS: multi-object tracking and segmentation. In: 2019 IEEE/CVF Conference on Computer Vision and Pattern Recognition (CVPR). IEEE, Long Beach, California, pp 7934–7943

80. Zhou X, Koltun V, Krahenbuhl P (2020) Tracking objects as points. In: 2020 European Conference on Computer Vision (ECCV). Springer, Berlin, pp 474–490

81. Cao J, Pang J, Weng X, Khirodkar R, Kitani K (2023) Observation-Centric SORT: rethinking SORT for robust multi-object tracking. In: 2023 IEEE/CVF Conference on Computer Vision and Pattern Recognition (CVPR). IEEE, Vancouver, BC, pp 9686–9696

82. Zhang Y, Sun P, Jiang Y, Yu D, Weng F, Yuan Z, Luo P, Liu W, Wang X (2022) ByteTrack: multi-object tracking by associating every detection box. In: 2022 European Conference on Computer Vision (ECCV). Springer, Tel-Aviv, pp 1–21

83. Maggiolino G, Ahmad A, Cao J, Kitani K (2023) Deep OC-Sort: multi-pedestrian tracking by adaptive re-identification. In: 2023 IEEE International Conference on Image Processing (ICIP). IEEE, Kuala Lumpur, pp 3025–3029

84. Chen L, Ai H, Zhuang Z, Shang C (2018) Real-time multiple people tracking with deeply learned candidate selection and person re-identification. In: 2018 IEEE International Conference on Multimedia and Expo (ICME). IEEE, San Diego, California, pp 1–6

85. Xiang Y, Alahi A, Savarese S (2015) Learning to track: online multi-object tracking by decision making. In: 2015 IEEE/CVF International Conference on Computer Vision (ICCV). IEEE, Santiago, pp 4705–4713

86. Yang M, Wu Y, Jia Y (2017) A hybrid data association framework for robust online multi-object tracking. IEEE Trans Image Process 26(12):5667–5679. https://doi.org/10.1109/TIP.2017.2745103
87. Milan A, Rezatofighi SH, Dick A, Reid I, Schindler K (2017) Online multi-target tracking using recurrent neural networks. In: 2017 Proceedings of the AAAI Conference on Artificial Intelligence (AAAI). AAAI, San Francisco, California, pp 4225–4232
88. Sun P, Cao J, Jiang Y, Zhang R, Xie E, Yuan Z, Wang C, Luo P (2021) TransTrack: multiple object tracking with transformer. arXiv:2012.15460, 2021
89. Xu Y, Ban Y, Delorme G, Gan C, Rus D, Alameda-Pineda X (2023) TransCenter: transformers with dense representations for multiple-object tracking. IEEE Trans Pattern Anal Mach Intell 45(6):7820–7835. https://doi.org/10.1109/TPAMI.2022.3225078
90. Zeng F, Dong B, Zhang Y, Wang T, Zhang X, Wei Y (2022) MOTR: end-to-end multiple-object tracking with transformer. In: 2022 European Conference on Computer Vision (ECCV). Springer, Tel-Aviv, pp 659–675
91. Wang J, Chen D, Wu Z, Luo C, Dai X, Yuan L, Jiang Y-G (2023) OmniTracker: unifying object tracking by tracking-with-detection. arXiv:2303.12079, 2023
92. Hyun J, Kang M, Wee D, Yeung D-Y (2023) Detection recovery in online multi-object tracking with sparse graph tracker. In: 2023 IEEE/CVF Winter Conference on Applications of Computer Vision (WACV). IEEE, Waikoloa, Hawaii, pp 4839–4848
93. Yang R, Zhang X, Wang G, Wu H (2023) Motion prior-based dual markov decision processes for multi-airplane tracking. In: 2023 Computer Vision and Machine Intelligence (CVMI). Springer, Singapore, pp 401–414
94. Bae S-H, Yoon K-J (2014) Robust online multi-object tracking based on tracklet confidence and online discriminative appearance learning. In: 2014 IEEE/CVF Conference on Computer Vision and Pattern Recognition (CVPR). IEEE, Columbus, OH, pp 1218–1225
95. Yang F, Choi W, Lin Y (2016) Exploit all the layers: fast and accurate CNN object detector with scale dependent pooling and cascaded rejection classifiers. In: 2016 IEEE/CVF Conference on Computer Vision and Pattern Recognition (CVPR). IEEE, Las Vegas, Nevada, pp 2129–2137
96. Li B, Yan J, Wu W, Zhu Z, Hu X (2018) High performance visual tracking with siamese region proposal network. In: 2018 IEEE/CVF Conference on Computer Vision and Pattern Recognition (CVPR). IEEE, Salt Lake City, Utah, pp 8971–8980
97. Bertinetto L, Valmadre J, Henriques JF, Vedaldi A, Torr PHS (2016) Fully-convolutional siamese networks for object tracking. In: 2016 European Conference on Computer Vision (ECCV). Springer, Amsterdam, pp 850–865
98. Afifi M, Brown MS (2020) Deep white-balance editing. In: 2020 IEEE/CVF Conference on Computer Vision and Pattern Recognition (CVPR). IEEE, Seattle, Washington, pp 1394–1403
99. Cheng J, Tsai Y-H, Wang S, Yang M-H (2017) SegFlow: Joint learning for video object segmentation and optical flow. In: 2017 IEEE/CVF International Conference on Computer Vision (ICCV). IEEE, Venice, pp 686–695
100. Afifi M, Price B, Cohen S, Brown MS (2019) When color constancy goes wrong: correcting improperly white-balanced images. In: 2019 IEEE/CVF Conference on Computer Vision and Pattern Recognition (CVPR). IEEE, Long Beach, California, pp 1535–1544
101. Guo C, Li C, Guo J, Loy CC, Hou J, Kwong S, Cong R (2020) Zero-reference deep curve estimation for low-light image enhancement. In: 2020 IEEE/CVF Conference on Computer Vision and Pattern Recognition (CVPR). IEEE, Seattle, Washington, pp 1777–1786
102. Wang R, Zhang Q, Fu C-W, Shen X, Zheng W-S, Jia J (2019) Underexposed photo enhancement using deep illumination estimation. In: 2019 IEEE/CVF Conference on Computer Vision and Pattern Recognition (CVPR). IEEE, Long Beach, California, pp 6842–6850
103. Xu K, Yang X, Yin B, Lau RWH (2020) Learning to restore low-light images via decomposition-and-enhancement. In: 2020 IEEE/CVF Conference on Computer Vision and Pattern Recognition (CVPR). IEEE, Seattle, Washington, pp 2278–2287

104. Afifi M, Derpanis KG, Ommer B, Brown MS (2021) Learning multi-scale photo exposure correction. In: 2021 IEEE/CVF Conference on Computer Vision and Pattern Recognition (CVPR). IEEE, Nashville, Tennessee, pp 9153–9163
105. Li B, Peng X, Wang Z, Xu J, Feng D (2017) AOD-Net: all-in-one dehazing network. In: 2017 IEEE/CVF International Conference on Computer Vision (ICCV). IEEE, Venice, pp 4780–4788
106. Zhang H, Patel VM (2018) Densely connected pyramid dehazing network. In: 2018 IEEE/CVF Conference on Computer Vision and Pattern Recognition (CVPR). IEEE, Salt Lake City, Utah, pp 3194–3203
107. Qin X, Wang Z, Bai Y, Xie X, Jia H (2020) FFA-Net: feature fusion attention network for single image dehazing. In: 2017 Proceedings of the AAAI Conference on Artificial Intelligence (AAAI). AAAI, New York, pp 11908–11915
108. Liu Y, Pan J, Ren J, Su Z (2019) Learning deep priors for image dehazing. In: 2019 IEEE/CVF International Conference on Computer Vision (ICCV). IEEE, Seoul, pp 2492–2500

# Chapter 4
# Applications

**Abstract** The AGVS dataset series and the development of computer vision algorithms for airport ground surveillance aim to create intelligent airport applications that enhance operational efficiency and safety. Common applications include airport augmented reality, visual conflict warnings, and visual docking guidance, all leveraging algorithms for segmentation, recognition, and tracking. In this chapter, we will use the Airport Panoramic Enhanced Surveillance system (APES), designed by our team, as a case study to illustrate the design methodology for intelligent airport applications. The APES system, collaboratively developed by the University of Electronic Science and Technology of China and the Second Research Institute of the Civil Aviation Administration of China, has been implemented in numerous Chinese airports.

## 4.1 Background and Introduction of APES

In the realm of international civil aviation, the concept of remote towers has revolutionized surveillance technology. Traditionally, air traffic controllers have monitored aircraft takeoffs, landings, taxiing, and docking through windows in tall towers at airports. However, as airports expand and passenger and cargo volumes continue to rise, this reliance on visual observation has revealed significant limitations, particularly in conditions of poor visibility or obstructed views. To address this issue, the Airport Panoramic Enhanced Surveillance (APES) system was developed, serving as an implementation of remote tower technology. The APES system provides controllers with an immersive panoramic view, allowing effective monitoring even from outside the tower.

The APES system's core functionality lies in presenting a comprehensive, unobstructed view of airport ground activities to controllers through the integration of multiple high-definition cameras, advanced image processing technology, and real-time data analysis. Unlike traditional single-view systems, APES captures video from various angles and synthesizes a continuous panoramic stream using image stitching technology. This allows controllers to oversee the entire airport operation from the control room, akin to standing at the tower's highest point, whether monitoring a busy runway, a large apron, or a complex taxiway network.

In China and many regions worldwide, the rapid growth of the aviation industry has led large- and medium-sized airports to face significant operational challenges. The increase in aircraft numbers, coupled with complex weather conditions and airport layout changes, has strained traditional visual surveillance methods. The APES system addresses these limitations by providing clear video displays in low-visibility conditions, ensuring controllers can accurately monitor each aircraft's position and movements, thereby enhancing safety and operational efficiency.

Additionally, the system features aircraft detection and tracking, panoramic display capabilities, and multipoint positioning, enabling controllers to access reliable information under diverse weather conditions. The remote push function facilitates true "remote tower" operations, allowing controllers to stay updated on airport activities whether they are in the air traffic control tower, apron tower, or a distant ground control center.

## 4.2  System Principle

The front-end components of the APES system primarily consist of high-definition cameras and outdoor shields. Data collected by these cameras is transmitted to a server via a network. Additionally, the system integrates data from existing airport systems, such as the airport vehicle monitoring system, flight planning system, Advanced Surface Movement Guidance and Control System (A-SMGCS), and ground monitoring radar. After data integration, the APES system processes the collected information through various methods, including image stitching, enhancement, aircraft detection and tracking, and behavior analysis. The processed data is then input into a streaming media server for storage and distribution, with the distributed data forwarded to user terminals.

User terminal configurations are flexible, allowing users to tailor them to their needs. They can opt for a suspended LCD splicing screen for large-scale situation displays or a 4K laser projection seamless splicing screen for a comprehensive integrated view. Through this design, the APES system ultimately provides air traffic controllers with real-time remote panoramic surveillance of the airport ground. An example of both the front-end and user-end components of the APES system is shown in Fig. 4.1.

### 4.2.1  System Implementation

The system implementation consists of both hardware and software components. The hardware includes multiple fixed high-definition network cameras and PTZ (pan-tilt-zoom) network cameras, each with a horizontal viewing angle of no less

**Fig. 4.1** An example of the front end and user end of an APES system

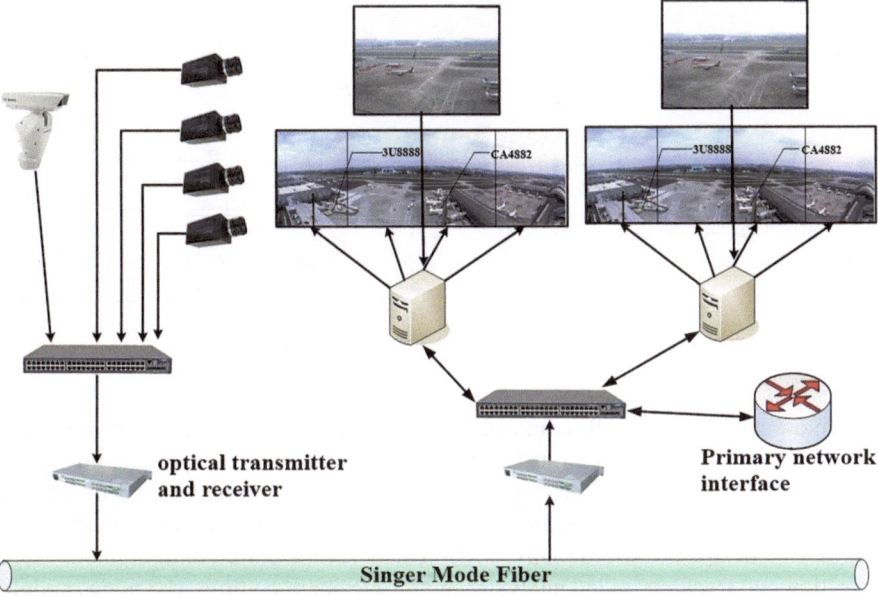

**Fig. 4.2** The hardware solution of the APES system. Adapted with permission from [1]

than 40°, installed throughout the airport. Additionally, network access equipment is used for video transmission, and high-performance computers equipped with processing software are deployed. In the monitoring and processing center, network equipment is necessary to receive video streams from various intelligent cameras, with DVR systems utilized for video storage. Furthermore, several high-performance workstations are required to perform tasks such as panoramic video stitching, image enhancement, video storage, multi-resolution target display, aircraft detection and tracking, coordinate positioning, and electronic map registration. The block diagram of the system's hardware technical solution is illustrated in Fig. 4.2.

In terms of software implementation, the system includes several key modules: the panoramic video stitching module, dynamic image enhancement module, multi-resolution display module, target detection and tracking module, and the electronic map coordinate mapping and trajectory processing module. Given that

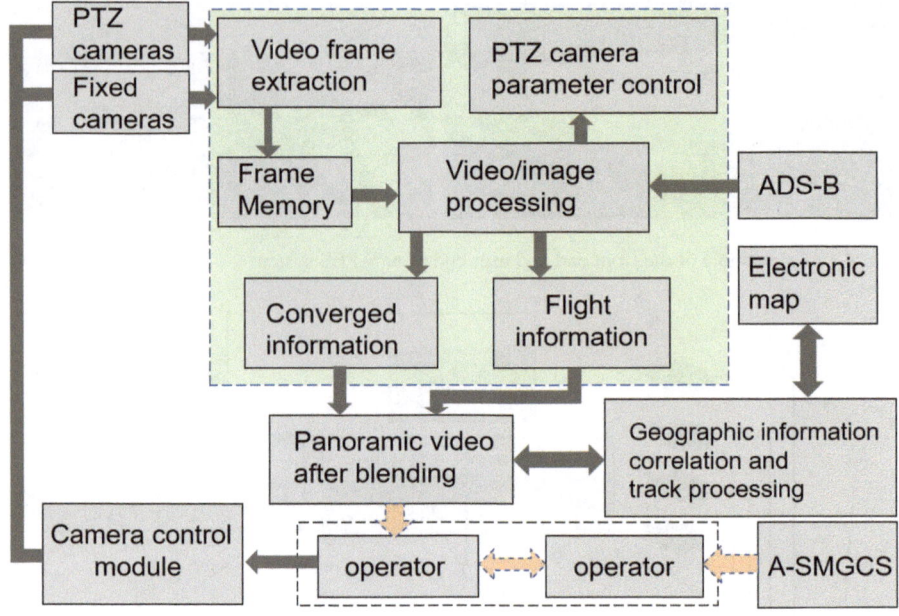

**Fig. 4.3** Schematic diagram of software implementation and data flow

the panoramic system handles multiple high-definition video streams and substantial data volumes, we employ a multi-CPU and GPU architecture for parallel processing to meet the system's real-time requirements. The CPU is tasked with multichannel control and video decoding, while the GPU handles panoramic video stitching and image enhancement. Additionally, a dedicated computer is configured for target detection and tracking, video analysis, and PTZ camera control. The main modules and data flows of the software component are illustrated in Fig. 4.3.

### 4.2.2  Multichannel Video Collection

The front end of the APES system consists of a camera array made up of multiple cameras. To achieve optimal panoramic video performance, careful planning is essential for the deployment of this camera array. The APES system utilizes the FOV-Crossed Cameras Setup method, where the Field of View (FOV) of each camera partially overlaps with that of its adjacent cameras. This deployment method is illustrated in Fig. 4.4.

In Fig. 4.4, the camera array is evenly distributed in a circular formation around a central point. Each camera captures a sector of the 360-degree Field of View (FOV), with slight overlap between the FOVs of adjacent cameras. This overlapping allows for a complete view of the 360-degree FOV when multiple cameras operate

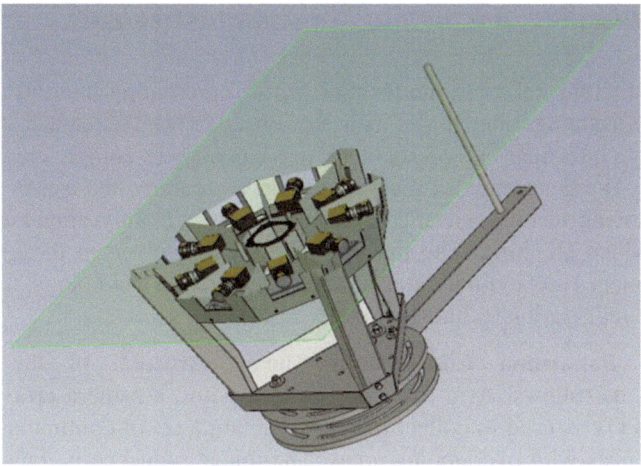

**Fig. 4.4** Illustration of the FOV-crossed cameras setup method

**Fig. 4.5** Illustration of the equipment protection cover

simultaneously. Although camera parallax can pose challenges in stitching videos from adjacent cameras, the vastness of the airport scene means that the distance from the cameras to the runway and aircraft is significant enough to minimize these parallax effects. Consequently, the deployment method depicted in Fig. 4.4 effectively supports airport ground surveillance. Additionally, since the area of interest for surveillance includes only the runway and apron, four cameras with a combined FOV of 180 degrees are sufficient. To address environmental factors such as temperature, humidity, and natural weather conditions like wind, snow, and rain, we designed a protection cover for the cameras to ensure their normal operation under these complex conditions. This protection cover is shown in Fig. 4.5.

### *4.2.3   Panoramic Stitching of Multichannel Videos*

Panoramic stitching refers to the technology of representing the 360-degree FOV observed by rotating one circle with the observer as the center. It is a new form of image information representation that can express complete environmental information. The generation of panoramic images requires the use of a variety of image processing technologies. The APES system is mainly aimed at horizontal panoramic video stitching. The main technologies involved in the generation of panoramic images are: cylindrical panorama construction technology, feature point-based image registration technology, etc.

**Cylindrical Panorama Construction** The construction of the cylindrical panorama is as follows. At a fixed observation point, a camera array covering a 360-degree FOV is used to collect images to obtain a set of continuous panoramic image sequences with overlapping areas. The image sequence is transformed into a cylindrical image sequence in a unified cylindrical coordinate space. Then, a suitable algorithm is used in the cylindrical space to splice adjacent cylindrical images to form a complete 360-degree cylindrical panoramic image. The principle of cylindrical projection transformation is shown in Fig. 4.6.

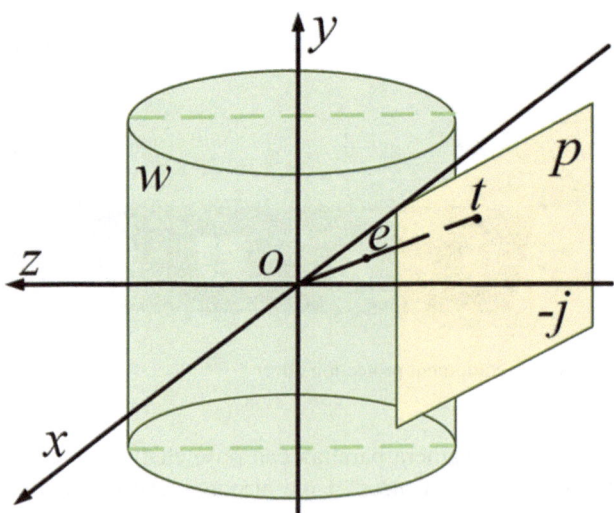

**Fig. 4.6** The principle of cylindrical projection transformation

$I$ is an image captured by a camera, and $p(x, y)$ is an arbitrary pixel in the image $I$. The projection transformation formula for projecting point $p$ onto the corresponding point $q(x, y)$ in cylindrical coordinate space is

$$\hat{x} = f \cdot \arctan((x - W/2)/f) + f \cdot \arctan(W/(2f)), \tag{4.1}$$

$$\hat{y} = f \cdot (y - H/2)/\sqrt{(x - W/2) + f^2} + H/2, \tag{4.2}$$

where $W$ and $H$ are the width and height of the image $I$, respectively, and $f$ is the radius of the projection cylinder (i.e., the focal length of the camera). For a 360-degree cylindrical panoramic image, the width of the image corresponds to the circumference $L$ of the cylinder, where $L = 2\pi f$. By estimating the width $L$ of the image, the initial focal length of the camera can be obtained as $f = L/2\pi$. When applying the cylindrical projection transformation to the image using the above formula, the transformation has the property that objects do not undergo distortion in the vertical direction. That is, two pixels $p1(x1, y1)$ and $p2(x2, y2)$ on the same vertical line in the image $I$ will still have the same horizontal coordinate in the cylindrical panoramic image. Using this property, each real-world image can be cylindrically projected, and then image stitching techniques can be applied to obtain a complete cylindrical panoramic image. An example of original images and cylindrical projection images is shown in Figs. 4.7 and 4.8.

**Image Stitching**  The core of image stitching lies in feature point extraction and registration. By registering the feature points in the overlapping areas of

**Fig. 4.7**  Original images to be stitched

**Fig. 4.8**  Cylindrical projection image

**Fig. 4.9**  Stitched image

adjacent images, the global coordinate correspondences between these images can be established. Once this is done, a simple coordinate transformation leads to the final stitched image. The APES system employs the Scale-Invariant Feature Transform (SIFT) for feature-based registration. The computation of SIFT features involves several key steps: constructing a difference of Gaussians scale space, detecting extrema, identifying stable feature points, and determining the main orientations of these points. After obtaining the feature points, image registration is performed to identify matching relationships between images. Since noise can introduce artifacts during feature point detection, noise removal is necessary. The system uses the Random Sample Consensus (RANSAC) algorithm to eliminate outliers during the matching process, improving accuracy. This method utilizes all measurement data, dividing them into inliers and outliers based on a set threshold. After discarding inaccurate outlier data, the more accurate inlier data is used for parameter estimation. The result of image stitching using this method for the example in Fig. 4.8 is shown in Fig. 4.9.

### 4.2.4   Image Enhancement

Since visible light imaging is highly sensitive to environmental factors like illumination, panoramic images stitched using the method from the previous section often exhibit noticeable stitching seams, with significant illumination differences in the overlap areas. To address this, image enhancement techniques are needed to achieve smoother transitions in these overlapping regions, improving the visual quality of the stitched images. Inspired by the concept of multiband blending, we propose an improved nonlinear weighted fusion algorithm. The core of this algorithm is to simulate the frequency domain scale represented by image texture details in the overlapping regions in a nonlinear manner. In terms of implementation, the algorithm first divides the overlapping region into three zones, with width ranges of $[3W/8, 5W/8]$, $[W/4, 3W/4]$, and $[0, W]$, where $W$ represents the width of the overlapping part. The division of these regions is illustrated in Fig. 4.10. The example shown in Fig. 4.9 demonstrates the result after applying image

**Fig. 4.10** Division of overlapping areas

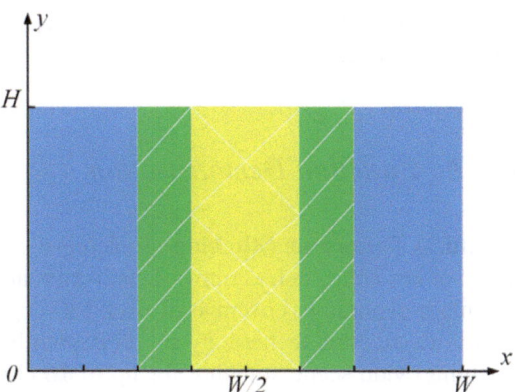

enhancement techniques. It is important to note that additional image enhancement methods are also integrated into the APES system, including nighttime image enhancement technology.

### 4.2.5   Multisource Information Fusion Based on Projection Transformation

There are multiple data sources in the APES system, including radar data, ADS-B data, etc., in addition to video data. Because video data is dominant, other data needs to be converted into the image coordinate before they can be used. This method of image coordinate conversion is called projection transformation. Projection transformation, such as the projection transformation from ADS-B data to image data, has been described in the previous chapter.

### 4.2.6   Target Detection and Tracking for Intelligent Applications

The APES system offers a range of intelligent airport application functions. The output from these applications is delivered to airport controllers through augmented reality (AR). AR enables the overlay of this output onto real-time panoramic video of the airport, allowing controllers to access additional operational information while monitoring the screen. All intelligent applications within the APES system are built on various computer vision algorithms, such as aircraft detection. Beyond detecting aircraft, the system analyzes both appearance and behavioral attributes of the targets, enabling real-time understanding and prediction of airport operations. The computer vision algorithms employed include segmentation, recognition, and tracking algorithms, along with several preprocessing techniques discussed in the previous chapter.

### 4.2.7   Function Demonstration

**Seamless Panorama Stitching**  To achieve comprehensive, no-blind-spot surveillance of key areas such as runways, taxiways, and aprons, the APES system provides seamless panoramic coverage of over 180 degrees, as illustrated in Fig. 4.11. Our real-time stitching technology supports both horizontal and vertical combinations, enabling wide scene stitching for up to 16 channels of high-definition video in a 2-row by 8-column configuration. This technology can process up to 25 frames per second. Through image preprocessing, precise feature point selection, matching, and image enhancement, the stitched monitoring images are guaranteed to be seamless, with no distortion, brightness discrepancies, or color inconsistencies, delivering a high-quality panoramic display for airport controllers.

**Airport Video Enhancement**  As previously mentioned, in addition to seam image enhancement, the APES system also provides other image and video enhancement functionalities. For instance, the system can effectively distinguish between buildings and moving objects on the ground, even in low-visibility conditions such as at night or in haze, as demonstrated in Figs. 4.12 and 4.13. The APES system intelligently suppresses glare and ensures automatic, evenly balanced exposure, thereby improving image quality. This is achieved using Wide Dynamic Range (WDR) technology, enabling precise handling of airport light halos. Additionally,

**Fig. 4.11**  Demonstration of seamless panoramic stitching performance

**Fig. 4.12**  Nighttime image before enhancement

**Fig. 4.13** Nighttime image after enhancement

**Fig. 4.14** The result of overlaying the flight number on the panoramic video

the system supports unified adjustment of brightness, color, and white balance across multiple cameras to ensure consistency in visual effects.

**Augmented Reality Demonstration** The APES system integrates multiple data sources, such as video data, radar data, A-SMGCS data, and ADS-B data, and incorporates a variety of video processing algorithms. These include image enhancement algorithms, image stitching algorithms, target detection and tracking algorithms, as well as various preprocessing and postprocessing techniques. Together, they enable a range of intelligent airport applications. Additionally, the APES system leverages augmented reality (AR) technology to overlay the output from these intelligent applications onto real-time video streams. For example, as shown in Fig. 4.14, flight numbers can be superimposed on the panoramic video display, providing an intuitive visual representation. The use of AR significantly enhances the clarity and real-time capabilities of airport ground surveillance, offering air traffic controllers a more efficient and user-friendly tool, ultimately improving air traffic management efficiency and safety.

# Reference

1. Zhang X, Qiao Y, Yang Y, Wang S (2023) SMod: Scene-Specific-Prior–Based Moving Object Detection for Airport Apron Surveillance Systems. IEEE Intell Transp Syst Mag 15(1):58–69. https://doi.org/10.1109/MITS.2021.3122926

# Chapter 5
# Conclusion and Future Directions

**Abstract** This book introduced the AGVS dataset series for airport ground video surveillance, multiple computer vision algorithms for this purpose, and a typical intelligent airport application. Future research will focus on enhancing computer vision algorithms for airport surveillance, exploring multimodal algorithms, designing new datasets, and developing innovative intelligent airport applications.

## 5.1  Summary of Key Findings

In the previous chapters, we systematically present the research status of airport ground video surveillance for the first time. We begin by outlining the background of civil aviation, highlighting the crucial role of airports in the industry and the significance of ground video surveillance within the concept of intelligent airports. Next, we detail research findings on airport ground video surveillance, focusing on three key areas: datasets, algorithm design, and application cases.

We introduced key datasets for airport ground video surveillance research, primarily the AGVS dataset series developed by our group. These datasets encompass segmentation, recognition, and tracking, supporting fundamental research in various aspects of airport video surveillance. Some AGVS datasets are publicly available, while others are being refined. Regarding algorithm research, we emphasized a unique design principle that leverages airport-specific prior knowledge to guide the development of computer vision algorithms. Our group has conducted preliminary investigations based on this principle, yielding useful results, some of which are detailed in the third chapter of this book. Finally, we showcased the Airport Panoramic Enhanced Surveillance (APES) system, an intelligent airport application from our group, as an example to illustrate the design method for such applications.

## 5.2   Future Research Directions in Airport Ground Video Surveillance

Future research directions for airport ground surveillance encompass four key areas: further exploration of the unique algorithm design principle developed by our group, investigation of multimodal algorithms, creation of new datasets, and development of new applications. We elaborate on each aspect below.

We proposed a design principle for computer vision algorithms in airport ground video surveillance, utilizing unique airport scene knowledge to guide algorithm development. Our research focused on segmentation, recognition, and tracking, yielding initial results. However, it remains preliminary, with limited prior knowledge used and a primary emphasis on segmentation. Moving forward, we aim to explore additional prior knowledge in airport scenes and conduct in-depth research across all three areas, along with other relevant technologies for intelligent airport applications, such as preprocessing and postprocessing.

Recent research on large models has seen remarkable advancements, particularly in multimodal applications, which typically involve text, speech, and images. While some of our computer vision algorithms, informed by prior knowledge of airport scenes, can be classified as multimodal, they differ significantly from those in large models. For instance, our ADS-B data-based computer vision algorithm requires converting ADS-B data into image format, making it fundamentally an image algorithm. In contrast, large models utilize joint learning to simultaneously capture features from multiple modalities, develop shared representations, and enhance interaction through techniques like contrastive learning, generative adversarial networks, and transfer learning. These approaches represent true multimodal algorithms. Consequently, we will focus our research on genuine multimodal algorithms for airport ground surveillance to facilitate the development of large airport models, starting with multimodal algorithms that integrate text and images.

The AGVS dataset series introduced in the second chapter consists of images and videos. Some datasets, like AGVS-T and AGVS-R, have been publicly released, while others, such as AGVS-AR, are still in production. Released datasets like AGVS-T have had multiple versions, including AGVS-T23 and AGVS-T24, and are continuously being improved. Additionally, to support research on multimodal algorithms for airport ground surveillance, we aim to create real multimodal datasets, such as those combining text and images. These new datasets and the algorithms based on them will be featured in the next edition of this book.

In the previous chapter, we introduced the APES system, an intelligent airport application designed by our group. We are also developing several other applications, including a visual conflict warning system and a visual docking guidance system. Additionally, we plan to create a large airport model and develop various intelligent applications based on it, which will be featured in the next edition of this book.